SUPERHYDROPHOBIC SURFACES - FABRICATIONS TO PRACTICAL APPLICATIONS

Edited by **Mehdi Khodaei, Xiuyong Chen** and **Hua Li**

Superhydrophobic Surfaces - Fabrications to Practical Applications
http://dx.doi.org/10.5772/intechopen.75218
Edited by Mehdi Khodaei, Xiuyong Chen and Hua Li

Contributors

Bornali Sarma, Colin Crick, Irena Roterman, Leszek Konieczny, Mateusz Banach, Mehdi Khodaei, Hua Li, Sepehr Shadmani, Xiuyong Chen

First published in London, United Kingdom, 2020 by IntechOpen
IntechOpen is the global imprint of INTECHOPEN LIMITED, registered in England and Wales, registration number: 11086078, 7th floor, 10 Lower Thames Street, London,
EC3R 6AF, United Kingdom
Printed in Croatia

British Library Cataloguing-in-Publication Data
A catalogue record for this book is available from the British Library

Additional hard and PDF copies can be obtained from orders@intechopen.com

Superhydrophobic Surfaces - Fabrications to Practical Applications, Edited by Mehdi Khodaei, Xiuyong Chen and Hua Li
p. cm.
Print ISBN 978-1-83880-597-5
Online ISBN 978-1-83880-598-2
eBook (PDF) ISBN 978-1-83880-605-7

For EU product safety concerns:
IN TECH d.o.o., Prolaz Marije Krucifikse Kozulić 3, 51000 Rijeka, Croatia,
info@intechopen.com or visit our website at intechopen.com.

Meet the editors

Mehdi Khodaei received his Bachelor's degree in 2005 and Master's degree in 2008 from the Department of Materials Engineering, Isfahan University of Technology, Iran. In 2014, he obtained his PhD degree from the University of Tehran, Iran in Nanotechnology-Materials Science. He was also a visiting researcher at Pohang University of Science and Technology (POSTECH), Korea, (March 2012–June 2013). His research area is in the field of Functional Nanocoatings, which is based on chemical, Sputtering deposition, pulsed laser deposition, and suspension thermal spray coatings. He is currently working as an assistant professor at the Faculty of Materials Science and Engineering, K.N. Toosi University of Technology, Iran.

Xiuyong Chen is an Associate Professor at the Ningbo Institute of Materials Technology and Engineering, Chinese Academy of Sciences, China. His current research is focused on thermal sprayed coatings, superhydrophobic surfaces, and cavitation erosion. In 2012, he achieved his PhD from the University of Chongqing, China. After having completed his PhD on biomedical engineering in 2012, he obtained a postdoctoral position at the Ningbo Institute of Materials Technology and Engineering (2013 - 2015), Chinese Academy of Sciences, China. He worked as a Visiting Professor at the University of Alberta, Canada (2018 - 2019). Dr. Chen has published 32 patents and over 30 peer-reviewed papers.

Hua Li is a Professor at the Ningbo Institute of Materials Technology and Engineering, Chinese Academy of Sciences, China. He has devoted over 20 years to fundamental and applied research on new coating materials, surface chemistry and physics, and design and construction of novel surface coatings. He received his Bachelor's degree in 1994 and Master's degree in 1997 from the Xi'an Jiaotong University, China. In 2002, he obtained his PhD degree from the Nanyang Technological University, Singapore. He worked as a SMF Research Fellow at the Nanyang Technological University, Singapore (2002 – 2006) and as a Research Associate at the Brookhaven National Laboratory, USA (2006 – 2010). Dr. Li has published 109 papers. He is the editor of the Journal of Thermal Spray Technology.

Contents

Preface

Superhydrophobicity was first observed in nature on a lotus leaf and some other plants when their leaves would not get wet. The main reason for that phenomenon was the unique surface structure of the lotus leaf and also the presence of a low surface energy material on the surface of the leaf. In order to achieve superhydrophobic surface or coating, the surface must possess hierarchical micro and nano roughness and low surface energy at the same time. Hierarchical micro and nano scale roughness will trap air on the surface that will cause an increase in the water contact angle and low surface energy will decrease the tendency of water to bond with the surface. So almost all the methods to achieve superhydrophobicity consist of two requirements of a hierarchical surface roughness as well as presence of a low surface energy material. These surfaces have many practical applications, from industrial to biomedical applications, including water/oil separation, self-cleaning, drag reduction, anti-fogging, anti-bacteria, anti-fouling, anti-icing, corrosion resistance, as well as many applications in industries such as marine, oil, and gas, aerospace, biomedicine etc. Hence, superhydrophobic surfaces, which can be achieved by surface modifications and/or surface coatings, have become very interesting in the last decade. The important issues and challenges in the field of superhydrophobic surfaces is stability and robustness of the surfaces.

Dr. Mehdi Khodaei
K.N. Toosi University of Technology
Tehran, Iran

Prof. Hua Li
Ningbo Institute of Industrial Technology (CNITECH)
Chinese Academy of Sciences (CAS)
Zhejiang, China

Dr. Xiuyong Chen
Ningbo Institute of Industrial Technology (CNITECH)
Chinese Academy of Sciences (CAS)
Zhejiang, China

Chapter 1

Introductory Chapter: Superhydrophobic Surfaces - Introduction and Applications

Mehdi Khodaei

Additional information is available at the end of the chapter

http://dx.doi.org/10.5772/intechopen.85359

1. Introduction

Superhydrophobicity was first observed in the nature on lotus leaf and in some other plants in which their leaves would not get wet. The main reason of this phenomenon was the unique surface structure of the lotus leaf and also presence of a low surface energy material on the surface of the leaf. In order to achieve superhydrophobic surface or coating, the surface must possess hierarchical micro- and nano-roughness and low surface energy at the same time. Hierarchical micro- and nanoscale roughness will trap air on the surface that will cause increase in water contact angle, and low surface energy will decrease the tendency of water to have bonding with the surface. So, almost all the methods to achieve superhydrophobicity consist of two steps: first to make a hierarchical surface roughness and then surface modification by a low surface energy solution of some materials like fatty acids, fluoroalkyl silanes, etc.

Atoms and molecules of liquid and solid have higher energy on the surface because there are few chemical bonds on the surface. This energy of surface atoms or surface molecules is known as the surface tension or the surface free energy. This energy is shown by γ and is equal to energy per unit area needed to build surface in constant temperature and pressure (J/m^2 or N/m). In case solid and liquid are in direct contact with each other, the surface energy will be lower in comparison to the situation in which these two are separated. The relation between surface energies and adhesion work is shown in Dupre equation.

$$W_{SL} = \gamma_{SA} + \gamma_{LA} - \gamma_{SL} \tag{1}$$

In this equation, W_{SL} is the adhesion work per unit area, γ_{SA} is the surface free energy between air and solid, γ_{LA} is the surface energy between air and liquid, and γ_{SL} is the surface free energy between liquid and solid.

When water droplet is placed on the surface of the solid, these two will reach equilibrium and water droplet makes a specific angle with the surface known as water contact angle (θ_0). The total energy can be calculated by the below equation:

$$E_{total} = \gamma_{LA}(A_{LA} + A_{SL}) - W_{SL} A_{SL} \tag{2}$$

In this equation, A_{LA} and A_{SL} are, respectively, liquid/air interface and liquid/solid interface. In this situation, regardless of gravitational potential energy and in constant volume and pressure in the equilibrium, dE_{total} is considered equal to zero.

$$\gamma_{LA}(d\,A_{LA} + d\,A_{SL}) - W_{SL}\,d\,A_{SL} = 0 \tag{3}$$

For a droplet with constant volume, θ_0 can be calculated by the equation below:

$$d\,A_{LA}/d\,A_{SL} = \cos\theta_0 \tag{4}$$

Then according to these equations, $\cos\theta_0$ can be calculated by the equation below known as the Young's equation:

$$\cos\theta_0 = (\gamma_{SA} - \gamma_{SL})/\gamma_{LA} \tag{5}$$

2. Wetting models

Several wetting models have been defined to calculate contact angle on the surface. The first wetting model is Young's equation that was just mentioned. This model does not consider surface roughness of the solid surface. The Young's equation is shown below:

$$\cos\theta = \frac{\gamma_{SG} - \gamma_{SL}}{\gamma_{LG}} \tag{6}$$

In this equation, θ is contact angle and γ_{SG}, γ_{SL}, and γ_{LG} are, respectively, surface free energy of solid/gas, solid/liquid, and liquid/gas interface.

It is obvious that in most cases, the surface is not smooth; so, Young's equation is not able to calculate the contact angle properly; thus, Wenzel equation was introduced. In this equation, it is considered that the surface wetting occurs uniformly and the equation is shown below:

$$\cos\theta w = r\cos\theta \tag{7}$$

In this equation, θw is the Wenzel contact angle, θ is Young's contact angle, and r represents the surface roughness factor that is equal to ratio of real surface to apparent surface.

As mentioned before, wetting is considered to be uniform in Wenzel's equation, or in other words, it is considered that water went through all surface cavities and there is no dry part. On the other hand, there is another wetting model that considers that the wetting is not

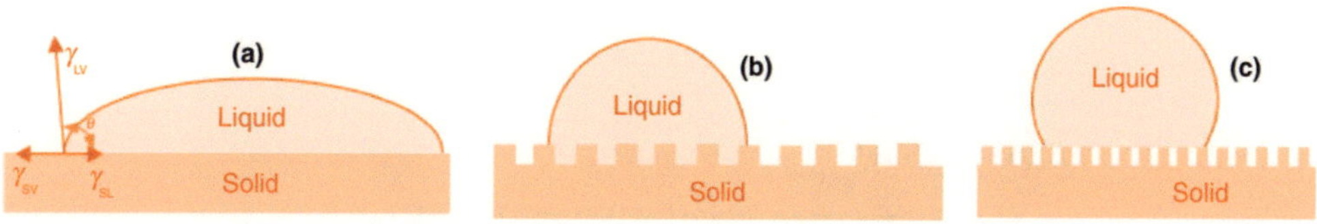

Figure 1. Schematic illustration of (a) Young's model, (b) Wenzel model, and (c) Cassie-Baxter model.

uniform and air packets do not let water to get into the surface cavities. In this case, water is in contact with solid and air packets, and water contact angle with air is equal to 180°. The model is called Cassie-Baxter and the equation is shown below:

$$cos\,\theta_{CB} = f_1 * cos\,\theta_1 + f_2 * cos\,\theta_2\; cos\,\theta_{CB} = f_1 * cos\,\theta_0 + f_2 * \cos(\pi) \tag{8}$$

$$cos\ \theta_{CB} = f_1 * cos\theta - f_2 \tag{9}$$

$$\cos\theta_{CB} = f_1 * (\cos\theta + 1) - 1 \tag{10}$$

In the above equations, θ_{CB} is the Cassie-Baxter contact angle, f_1 is the ratio of area that liquid is in contact with solid, and f_2 is that ratio of area that liquid is in contact with air packets made or trapped air inside the surface cavities. In **Figure 1**, the difference between three aforementioned wetting models is shown [1].

3. Application of surfaces with superhydrophobic properties

Superhydrophobic surfaces and coatings as mentioned have a unique behavior against water droplets. This unique behavior result into a new set of applications including self-cleaning, anti-icing, antibacterial, oil-water separation, corrosion resistance, etc. Some applications are described below.

3.1. Oil-water separation

There have been many reports of oil contaminants in sea waters and rivers due to leak of factories waste into nature and accidents like Deep Water Horizon. Removing oil contaminants from water was always challenging and expensive; so, different methods have been introduced by scientists in order to remove them. These methods are categorized into three main groups: water removing, oil removing, and smart controllable separators. The water removing filters are superhydrophilic and superoleophobic; this kind of filters works under water and when they get wet by water, the presence of the water on the surface of the filter prevents oil to pass from the filter pores. The category in oil removing method which by my personal opinion is a more efficient way because the amount of oil is always less than the amount of water; so, it is logical that we try to remove oil from water and not water from oil. To remove oil from

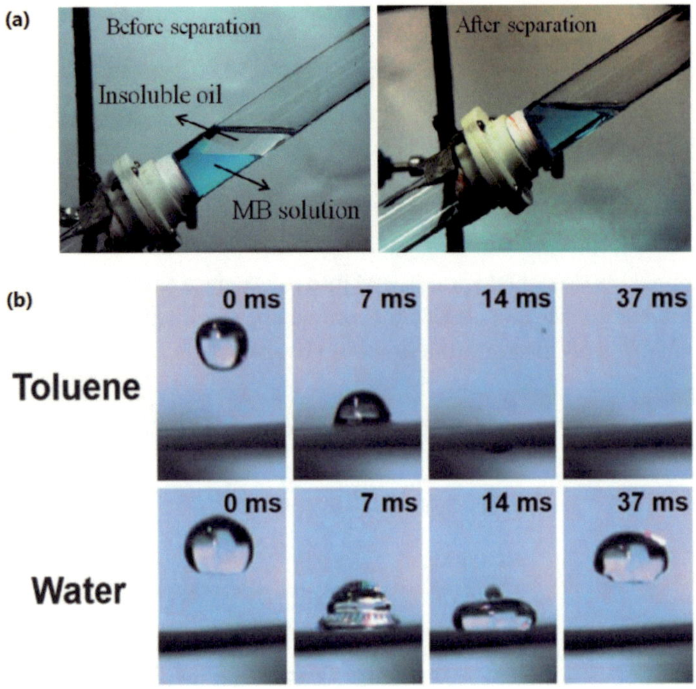

Figure 2. (a) Oil–water separation with use of TiO_2-coated superhydrophobic and superoleophilic mesh [2], (b) opposite behavior of silicone elastomer-coated mesh against water and toluene droplets [3].

water, the material should be superhydrophobic and superoleophilic. Superhydrophobic oil removing filters are the main part of the oil removing category. Gao et al. [2] used a TiO_2-coated mesh in order to separate oil from water (**Figure 2a**), and Crick et al. [3] used a silicon elastomer coating on a mesh to efficiently separate organic solvents like hexane, petroleum ether, and toluene from water. As shown in **Figure 2b**, water droplet cannot pass through the filter but toluene can easily pass through.

3.2. Corrosion resistance surfaces

There are several ways to protect a surface from corrosion. During the past two decades, scientists have been using superhydrophobic nanocomposite coatings without any toxic materials in order to protect various surfaces from corrosion. The corrosion protection capability of the superhydrophobic coatings mainly is because of the presence of air packets between surface and corrosive solution, and these packets act like a barrier and prevent from corrosive ions diffusion and protect the substrate [1].

Superhydrophobic metallic surfaces could be able to decrease the corrosion rate of metals by several orders of magnitude through imparting hydrophobization. Several reports have been published that demonstrated the enormous capability of superhydrophobic surfaces on the corrosion mitigation. The potentiodynamic polarization test revealed a significant decrease in the corrosion current density (**Figure 3**) of metallic surfaces by using a commercial hydrophobic surface modification [4].

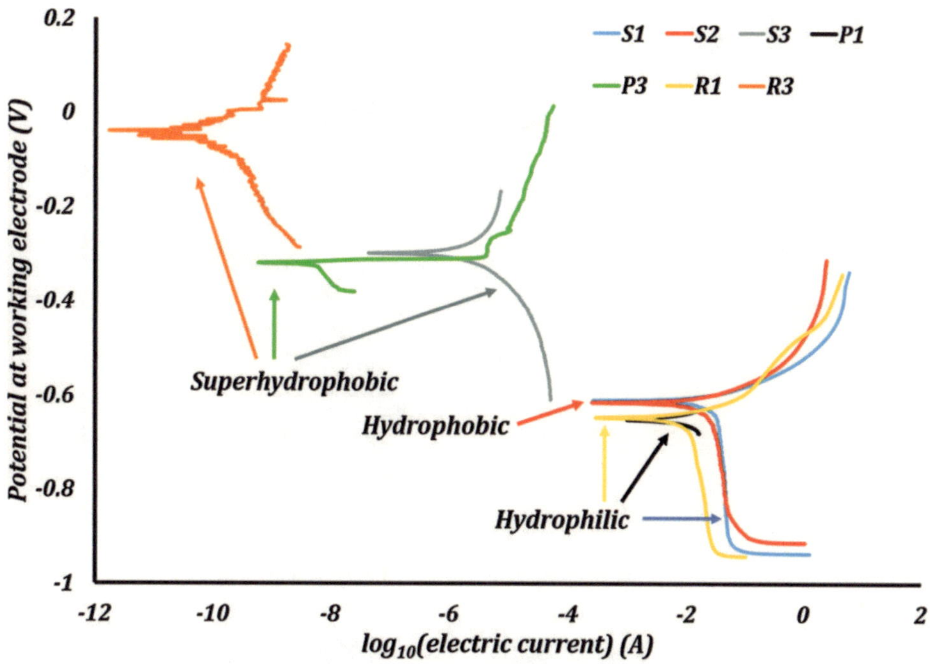

Figure 3. Potentiodynamic polarization curves of bare metallic surfaces (hydrophilic) and surface modified samples with a commercial hydrophobic material (hydrophobic) and with developed commercial hydrophobic materials (superhydrophobic) [4].

3.3. Self-cleaning properties

The lotus leaf's surface is always clean regardless of any contamination that may be present in its surrounding environment. This leaf has a unique surface structure coated with wax and shows superhydrophobic properties, and sliding angle is very low so water can easily slide on the surface of the leaf and remove any contaminants. The aforementioned properties of superhydrophobic surfaces and coatings are called self-cleaning properties. There are many superhydrophobic coatings which were synthesized with different methods and used in industries. It is worth to mention that the actual self-cleaning surface is the surface exhibiting the combined superhydrophilicity and photocatalytic behaviors to decompose the dirt. The use of the term, self-cleaning surface, is not appropriate for superhydrophobic surfaces, which are extremely dry and repel water drops. As schematically shown in **Figure 4**, these surfaces do not actually clean themselves but they wash away the dirt when the water drops roll over the surface.

3.4. Anti-icing properties

In recent years, superhydrophobic coatings have been suggested as anti-icing coatings. As mentioned before, the presence of air packets on the superhydrophobic surfaces causes the water droplets to slide easily on the surface; therefore, there will not be enough time for the droplet to freeze on the surface; consequently, this reduces the side effects of frosts on the surfaces.

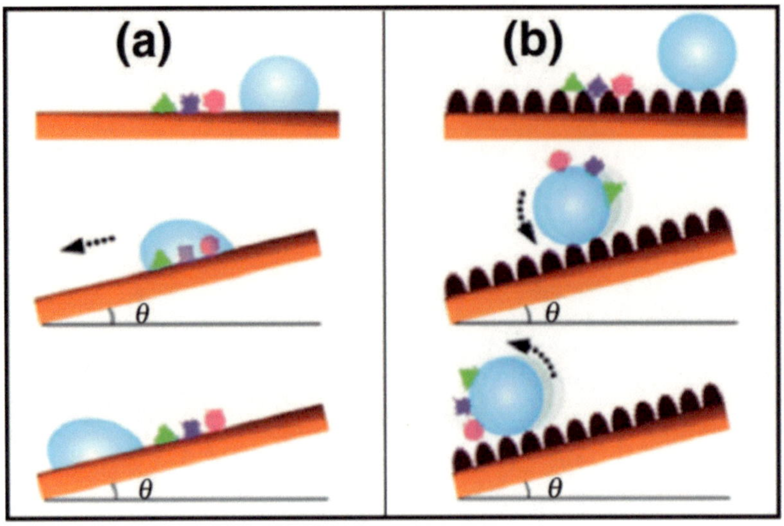

Figure 4. Schematic illustration of self-cleaning process in (a) non-hydrophobic and (b) hydrophobic surfaces.

Figure 5. Comparison of the ice formation on the uncoated and superhydrophobic-coated insulators [5].

Every year ice storms harm the equipment such as electrical transmission equipment, communication systems, aerospace facilities, highways, etc. In order to reduce this kind of damages, different methods have been developed such as local warming and preventing of ice formation by chemical activities and additives, which have some limitations in practical applications. On the other hand, the preventing of surface from ice development by superhydrophobicity phenomena could be practical in most cases without requiring special requirements and devices. One of the important applications of icephobic surfaces is using the insulators of transmission lines, which are needed to prevent the ice formation in a cold area. The experimental survey of ice formation on coated and uncoated surfaces of an insulator (**Figure 5**) under a condensing weather condition at −5°C and saturated humidity revealed that the superhydrophobic surface is completely effective in reducing ice adhesion to the surface up to 97%.

Actually scientists have some disagreements about the relations between superhydrophobicity and anti-icing properties. Some believe that these two are not related to each other; on the other hand, some insist that superhydrophobicity results to anti-icing properties. This disagreement is because there is no specific standard that can be used to evaluate ice adhesion to the surfaces; also, the method of preparing ice for each study is different from the others; so by now, it is not possible to have a definite answer to this matter. The recent studies have helped to get a better understanding about ice formation process on superhydrophobic surface but there is still much left unknown about the nucleation, growth, and adhesion to the surface which need more studies and information in this subject.

3.5. Drag reduction

Drag force is one of the major problems that a solid moving in water such as a ship or submarine faces. This force is resulted from the friction force between water and moving solid surface in the water. Inspired from shark skin, several superhydrophobic coatings were fabricated in order to reduce the drag. As mentioned before, superhydrophobic coatings have some air pockets inside their hierarchical micro- and nanoscale surface structures which will reduce the contact between solid and liquid so that the drag force will dramatically reduce. Drag reduction phenomenon by superhydrophobic surfaces was investigated in various works such as the one reported by Dong et al. [6], where they have fabricated a superhydrophobic coating on a model ship with a large and curved surface by electroless deposition of gold aggregates. The superhydrophobic model ship exhibited a remarkable drag reduction of 38.5% (**Figure 6**). On a non-coated sample, the friction is just between solid and water, but on a superhydrophobic

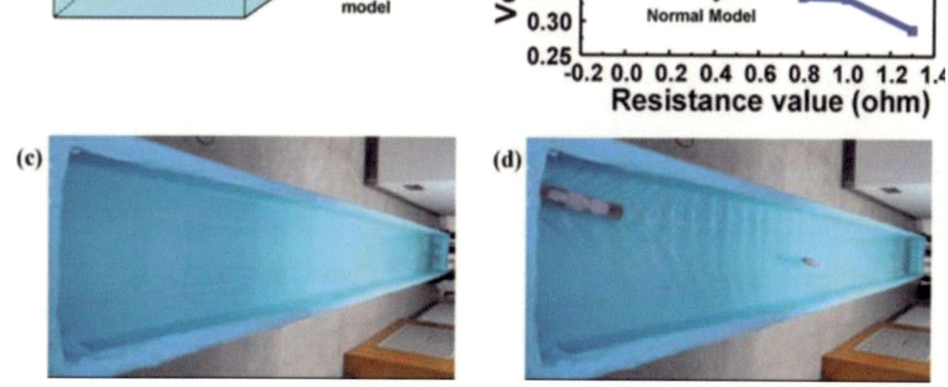

Figure 6. (a) Illustration of the device for the drag-reducing test; (b) velocity of the model ships with and without a superhydrophobic coating versus the values of the resistance in the circuit within the ship; snapshots of (c) at the beginning and (d) at the end of the drag-reducing test [6].

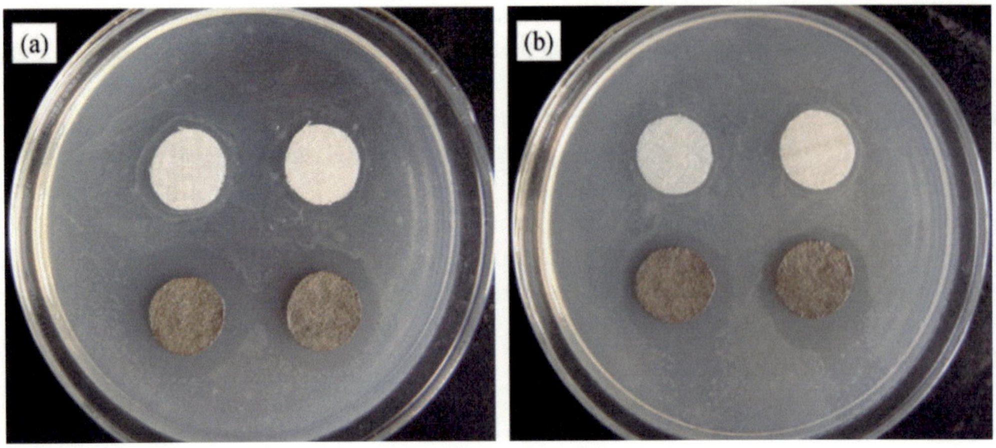

Figure 7. Antibacterial activity of (a) normal cotton (the upper two) and Ag NP modified cotton (the lower two) textiles, and (b) normal cotton (the upper two) and hydrophobized Ag NP modified cotton (the lower two) textiles [7].

surface, there are three phases, water, solid, and trapped air between these two; so, the friction will be drastically reduced in this situation which is known as the plastron effect.

3.6. Antibacterial properties

Antibacterial properties are essential in biosensors, implants, food packaging, and industrial and marine equipment. For example, one of the main reasons that cause infection in patient after surgery is bacteria that grow on implants. In order to solve this problem, antibacterial coatings that reduce the bacterial adhesion to the surface suitable are used. One research in this regard fabricated the silver nanoparticles on cotton fibers and then modified by the hexadecyltrimethoxysilane to get superhydrophobicity [7]. Antibacterial activity of the samples (inhibition zone formed on agar medium) has been determined as shown in **Figure 7**. The results showed that the normal cotton samples, exhibit no antibacterial activity, whereas the silver modified cotton surfaces killed all the bacteria under and around them showing a distinct inhibition zone with an average width of 8.78 mm around the samples.

Author details

Mehdi Khodaei

Address all correspondence to: mehdi.khodaii@gmail.com; khodaei@kntu.ac.ir

Faculty of Materials Science and Engineering, K. N. Toosi University of Technology, Tehran, Iran

References

[1] Jeevahan J, Chandrasekaran M, Joseph GB, Durairaj RB, Mageshwaran G. Superhydrophobic surfaces: a review on fundamentals, applications, and challenges. Journal of Coating Technology and Research. 2018;**15**:231-250

[2] Gao C, Sun Z, Li K, Chen Y, Cao Y, Zhang S, et al. Integrated oil separation and water purification by a double-layer TiO_2-based mesh. Energy & Environmental Science. 2013;**6**: 1147-1151

[3] Crick CR, Gibbins JA, Parkin IP. Superhydrophobic polymer-coated copper-mesh; membranes for highly efficient oil–water separation. Journal of Materials Chemistry A. 2013;**1**:5943-5948

[4] Ramachandran R, Nosonovsky M. Coupling of surface energy with electric potential makes superhydrophobic surfaces corrosion resistant. Physical Chemistry Chemical Physics. 2015;**17**:24988-24997

[5] Zhu L, Xue J, Wang YY, Chen QM, Ding JF, Wang QJ. Ice-phobic coatings based on silicon-oil-infused polydimethylsiloxane. ACS Applied Materials & Interfaces. 2013;**5**:4053-4062

[6] Dong H, Cheng M, Zhang Y, Wei H, Shi F. Extraordinary drag-reducing effect of a superhydrophobic coating on a macroscopic model ship at high speed. Journal of Materials Chemistry A. 2013;**1**:5886-5891

[7] Xue C-H, Chen J, Yin W, Jia S-T, Ma J-Z. Superhydrophobic conductive textiles with antibacterial property by coating fibers with silver nanoparticles. Applied Surface Science. 2012;**258**:2468-2472

Chapter 2

Approaches for Evaluating and Engineering Resilient Superhydrophobic Materials

Colin R. Crick

Additional information is available at the end of the chapter

http://dx.doi.org/10.5772/intechopen.80746

Abstract

Superhydrophobic materials rely upon highly rough surface morphologies in order to maximise water repellency, and requires surface features on the micro/nanoscale. These tremendously small surface structures are inherently physically weak, relative to characteristics of bulk materials. This limits the real-world applicability of many superhydrophobic surfaces, as degradation and loss of superhydrophobicity readily occurs upon exposure to anticipated stimuli. Consequently, there is an absence of long-lasting commercial products, but instead rely upon frequent regeneration. These materials demonstrate a tremendous potential for application in a range of areas, including antifouling, self-cleaning, drag-reduction, anti-icing, etc. To realise application on these fields, superhydrophobic resilience must be maximised. This chapter summarises evaluation methods and engineering procedures in attaining resilience, both are highly important in the development of robust materials.

Keywords: superhydrophobic, resilience, evaluation, engineering, degradation, microstructure, chemistry

1. Introduction

Superhydrophobic materials exhibit potential real-world application that encompasses a wide range of commercial sectors [1–3]. This is a result of properties inherent to superhydrophobic surfaces, including; self-cleaning, antibiofouling, drag-reduction, and oil-water separation. State-of-the-art research provides a tremendous breadth of superhydrophobic coatings and membranes reported within the literature [3]. However, there is a noticeable absence of commercial solutions currently available. The majority of superhydrophobic products concentrate on short-term treatments, which require reapplication of the coatings to retain functionality [4, 5].

The short-term nature of many applied superhydrophobic surfaces stem from the architecture required to induce extremely water repellent properties. The two major features of these materials are; (i) an inherently water repellent surface chemistry, and (ii) a highly rough surface microstructure [6–8]. The latter of these necessitates a surface structure composed of micrometre, or nanometre, sized features—which are fundamentally physically weak structures. Therefore, commercial products that impart superhydrophobicity tend to degrade overtime, deteriorating at a faster rate as the intensity of the application increases [9].

The major challenge facing researchers aiming for applicable materials is surface fragility, and routes for engineering resilience. This chapter aims details the nature of superhydrophobic degradation, and monitoring techniques, in addition to required materials tolerances, and a summary of approaches to achieving resilience.

2. Superhydrophobic surface degradation

2.1. Chemical/physical degradation

The loss of superhydrophobic properties can stem from changes to surface chemistry, or the loss of tapped air at a surface (e.g. by degradation of surface roughness), or a combination of these factors [10, 11]. Understanding the cause of the loss of superhydrophobicity enables the targeting of specialised surface design for resilience enhancement.

2.1.1. Surface chemistry changes

The surface chemistry of a material determines its Young contact angle; this value is a measure of wetting for a flat ideal surface [12]. The Young contact angle feeds into the main surface wetting models; Wenzel, and Cassie-Baxter [13, 14]. As most superhydrophobic materials are fabricated with an inherently water repellent coating, any deviation from this will lead to an observed reduction in water contact angle (WCA—**Figure 1**) [6]. This is demonstrated by the covalent attachment of self-assembled monolayers (SAMs), where the strength of SAM-surface binding is an extremely important consideration with respect to coating longevity. Another factor that affects the stability of surface bound molecules is molecular size, whereby variation in the size of these molecules can either enhance, or impair their relative stabilities [15]. Chemical robustness can be ensured through the utilisation of energetically stable surface chemistry, or the incorporation of hydrophobic coatings that exceed the thickness of a monolayer [16, 17].

Another pathway that includes surface chemical change is surface fouling, especially when considering real-world application. This issue can be seen when biofouling, or chemical contamination (e.g. hard water staining) is a concern [18, 19]. Surface fouling can be controlled through solutions to particular contaminants, however the success of a particular approach is very much dependent on the operational environment of the materials [20].

2.1.2. Physical degradation

A lowering in WCA achieved by a superhydrophobic surface is anticipated when the overall surface roughness is reduced [1–3]. However, a WCA reduction would be also expected

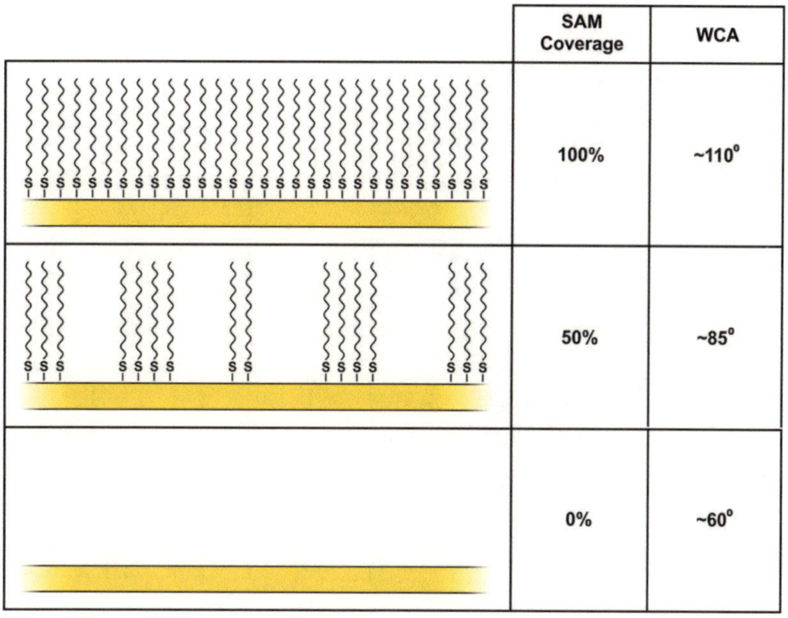

Figure 1. A demonstration of the effect of SAM surface coverage on the resultant WCA. This theoretical example utilises ideal Youngs WCAs for the complete SAM coating, and bare substrate of 110°, and 60° respectively [12–14].

through any unfavourable alteration of the water-surface wetting interface (e.g. loss of trapped air) [21]. This can be reasoned through examining the roughness terms in the Cassie-Baxter surface model, where f_1 represents the liquid-solid interfacial unit area (akin to the 'roughness factor' within the Wenzel model), and where f_2 represents the liquid-air interfacial unit area (a 'roughness factor' for the air trapped at the interface). Therefore, any change to the solid roughness (e.g. surface microstructure damage) is reflected by a reduction f_1, and the removal of air causing a reduction in f_2 (**Figure 2**) [14].

The removal of trapped air, without any additional surface variation, can be considered a form of recoverable degradation (covered in Section 2.2), if this air is allowed to re-enter the surface porosity [22]. However, recovery from physical damage (i.e. scratching, material removal, or flattening) cannot be achieved simply, unless engineered into the surface components (e.g. self-healing materials—Section 4.3). Resilience to this type of damage requires is a key engineering challenge within superhydrophobic research (detailed in Section 4).

2.2. Recoverable degradation

As outlined in the previous section (Section 2.1.2), not all degradation is results in a permanent change in wetting behaviour [22]. The removal of trapped air, transitioning from Cassie-Baxter to Wenzel type wetting, can result in a change in the way water interacts with the surface, in addition to the loss of superhydrophobicity (**Figure 2(iii)/3**) [23]. This air can be removed physically (*via* hydrostatics), additionally it can also be slowly dissolved by water over time [24, 25]. Hydrostatic removal can occur when water interacts dynamically with the surface (e.g. water impact, turbulent surface flow, etc.), whereas air solvation most commonly

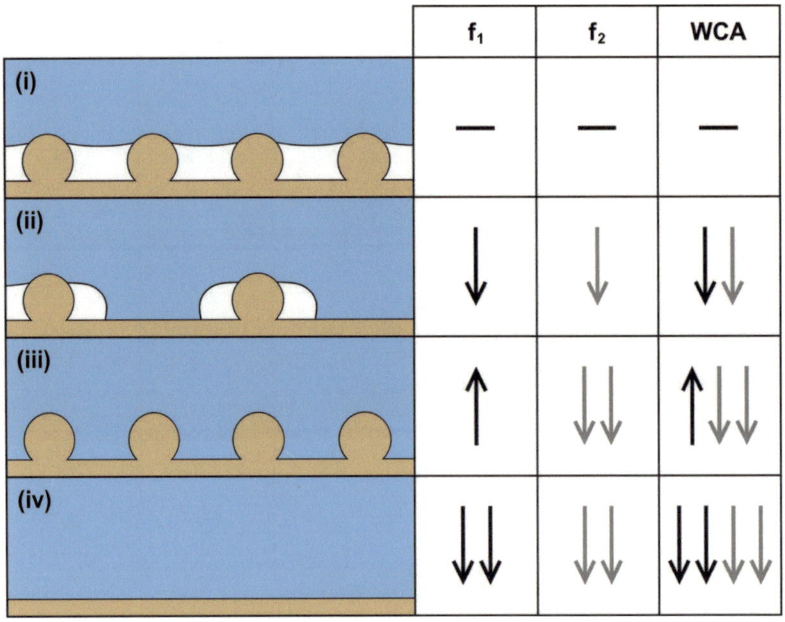

Figure 2. Scheme showing the effect of variation of surface roughness and trapped air (f_1/f_2) on the on the WCA for a surface. Where; (i) is the original superhydrophobic surface wetting (i.e. Cassie-Baxter), (ii) shows the partial removal of roughness features, (iii) demonstrates the Cassie-Baxter to Wenzel transition (*NB*; f_1 increases), and (iv) complete removal of all surface features. The schematic presumes a consistent surface chemistry [12–14].

occurs when superhydrophobic materials are submerged underwater for extended periods. In either case, once air is reintroduced and the Cassie-Baxter state is regained, surface hydrophobicity (i.e. WCA, hysteresis, etc.) will return to the original state (**Figure 3**) [22].

2.3. Degradation analysis

The optimisation of a materials resilience can be facilitated through understanding the nature of superhydrophobic degradation (changes to; surface chemistry, microstructural damage, or recoverable degradation), within the context of the type of resilience testing used, or degradation stimuli applied [26]. The following section details commonly used degradation, and degradation analysis protocols.

2.3.1. Degradation protocols

A commonly used approach in testing the physical robustness of superhydrophobic materials includes an array of scratch or abrasion testing methods [26]. In the literature, this ranges from arbitrary scratching of the surface with a blade or scalpel, to quantitative abrasion testing compatible with industrial standards (e.g. linear abrading) [26–28]. Arbitrary testing methods (e.g. blade scratching) can provide a good indication of surface resilience, particularly if the exact specifications of testing are reported (**Figure 4**). However, many examples throughout the literature do not provide adequate detail to assess materials resilience (e.g. blade testing; blade type, blade dimensions, scratch protocol—force applied/blade travel/etc., or indication

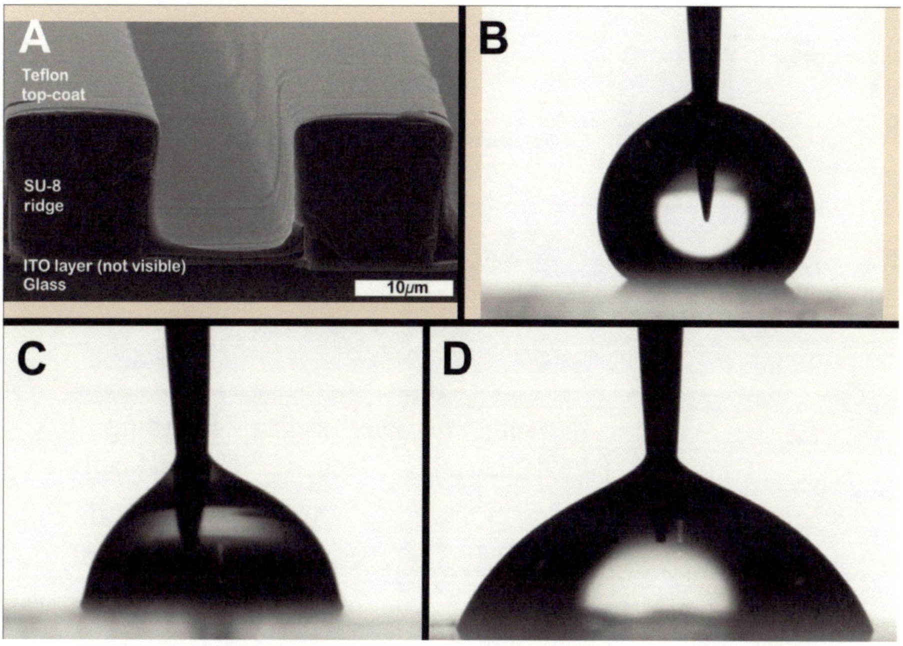

Figure 3. SEM/photos of substrates fabricated to probe the Cassie-Baxter to Wenzel wetting transition, and their wetting behaviour. The ridge surface structures (A) facilitate superhydrophobicity *via* the trapping of air (B). Once fully wetted, the surface demonstrates a direction dependent WCA variation (C/D). The original WCA are regained when samples are dried and retested.

of reproducibility) [27]. Another commonly used method employs adhesive tape, which is applied to the superhydrophobic material and then removed to cause surface damage. Although this may provide some insight into surface resilience, the techniques are associated with many variables (e.g. adhesive strength, tape application/removal protocols, surface microstructure, etc.) imparting a level of uncertainty to the testing. In contrast, quantitative abrasion tools are able to provide a reproducible method of surface degradation (**Figure 4**). While providing an objective indication of the magnitude of surface roughness. The utilisation of a commercial linear abrader has been shown to provide details of; abradant type/size/contact area, and abrasion force/velocity [28].

These testing examples focus in the physical degradation of the surface material (i.e. the flattening of surface roughness), subsequently affecting surface hydrophobicity. Recoverable degradation (without surface damage) is less often probed, however several literature protocols have been developed [30]. One example of probing the Cassie-Baxter to Wenzel transition, is the use of water-surface compression apparatus. This utilises a parallel plate configuration, where water is place between two superhydrophobic surfaces which are then progressively moved closer together (**Figure 5**). The result is a surface tension induced pressure increase at the interface, which progressively squeezes air from the surface microstructure (hence transitioning from Cassie-Baxter to Wenzel wetting). The pressure required for the wetting transition is a measure of the stability of the superhydrophobic trapped air stability [31]. An example aimed at accessibility uses water bouncing to characterise superhydrophobic

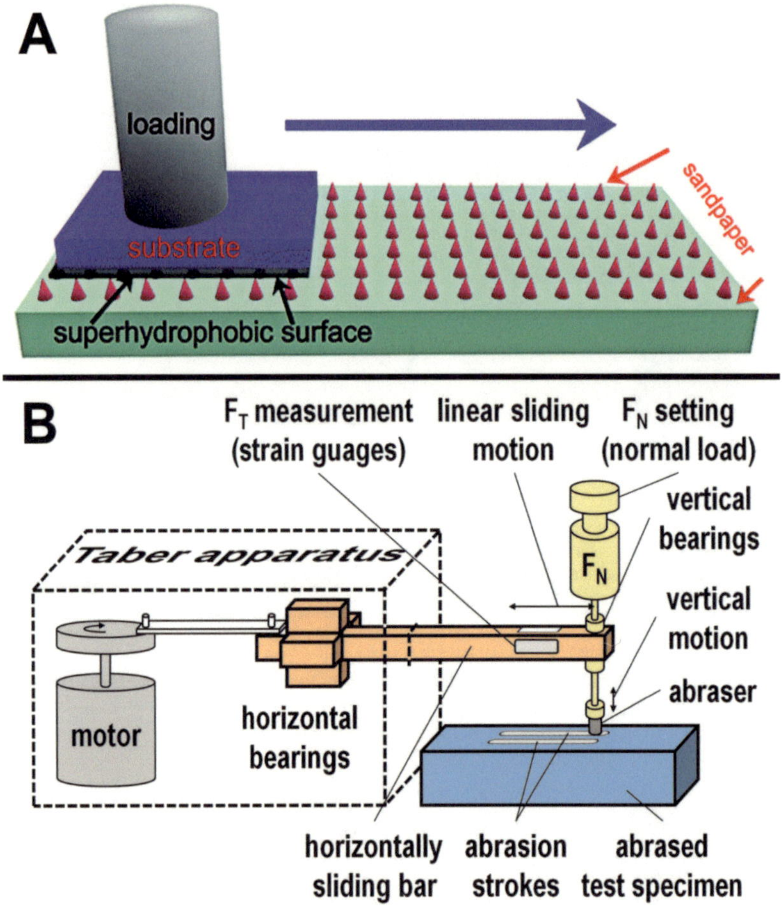

Figure 4. Physical degradation protocols utilising arbitrary surface abrasion (A—weighted abrasion cycling), and quantitative industrial abrasion (B—linear abrader) [27, 29].

resilience. The technique enables an assessment of trapped air stability, however does not require the use of specialised equipment (e.g. parallel plate set-up) [32, 33].

The use of dynamic water interactions with superhydrophobic surfaces has been employed to examine both physical degradation, and the reversible removal of air at the water-solid interface. This includes water flow, and water-surface collisions, and ranges from arbitrarily flowing water across a surface, to quantitative techniques (e.g. sheer stress endurance) [34–37]. Water sheer has been shown to be provide the conditions for testing reversible and non-reversible degradation, while offering in situ monitoring of surface hydrophobicity (detailed in Section 2.3.2—**Figure 5**). The application of water sheer stress provides a direct measurement of the materials resilience to water flow across the surface [34]. In contrast, the wider relevance to other forms of degradation (e.g. scratch resistance) can only be considered with some amount of ambiguity [27].

The development of highly resilient superhydrophobic materials is a key concern within the research field [1–3]. Tremendous progress has been made in the engineering of materials

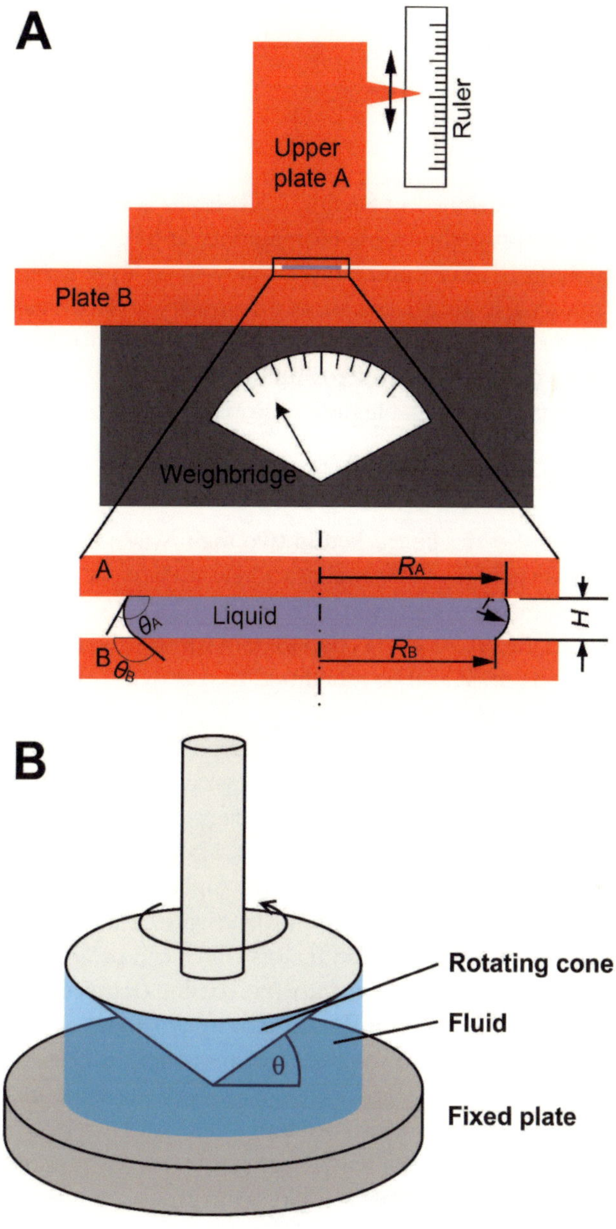

Figure 5. Recoverable degradation testing, *via* parallel plate pressuriser (A), and cone-on-plate rheometry (B) [31, 34].

robust enough to survive real-world application (summarised in Section 4). However, the lack of standardised testing, through the use of arbitrary (e.g. scalpel scratch testing), or semi-quantitative (e.g. abrasive drag testing) techniques, hinders potential advancement, as this leaves some literature open to interpretation. As mentioned within this section, quantitative techniques are able to provide an accurate indication of materials resilience to a particular stimulus. In many cases, further analysis is required to provide a comprehensive indication of robustness. This ambiguity pushes the development of a standardised testing regime, which takes into account a range application-specific testing protocols, particularly if a material is targeted at broad real-world application.

2.3.2. Degradation analysis protocols

The degradation pathways (detailed in Section 2.3.1) inflict a range of chemical, structural, and recoverable changes on superhydrophobic materials [26]. For physical/chemical changes, post degradation analysis is most commonly employed, as in situ monitoring of surface roughness (macro/nanoscale features), and surface chemistry is not trivial [38–40]. The reversible removal of air can be successfully tracked using in situ monitoring, as it is suitable to the dynamic wetting/drying of the surface microstructure [41]. The following section highlights the range of analysis techniques available for monitoring these types of degradation:

2.3.2.1. Surface chemistry alteration

Changes to surface chemistry can be tracked in two main ways; (i) analyse the result of the changing surface chemistry, or (ii) directly characterise surface chemistry [42, 43]. Primarily, WCAs can be used to estimate the variation in hydrophobicity [42]. This is relatively straightforward, however due to the WCAs connection to both surface chemistry, and surface roughness, morphological effects must also be monitored to specify the nature of chemical change [10]. Direct characterisation of surface chemistry can be conducted using a multitude of techniques [43]. Vibrational spectroscopies (e.g. infrared/Raman) provide a non-destructive indication of surface functionality and are commonly employed. Monitoring of surface chemistry is important particularly when environmental degradation pathways are likely, such as superhydrophobic materials constructed from photocatalytic materials (e.g. Titanium dioxide, TiO_2). TiO_2 nanoparticles are commonly used for superhydrophobic coatings, as they are commercially applicable (white pigment utilised worldwide) and can be easily functionalised to induce hydrophobicity. TiO_2 is a semiconductor, which can produce highly active species which break down organic species on the surface (i.e. coatings required for superhydrophobic materials) [44]. Therefore, a reduction in the amount of hydrophobic surface coating can be tracked using infrared, in addition to the resultant reduction in observed WCA (**Figure 6**) [45, 46]. X-ray photoelectron spectroscopy is a highly sensitive technique, able to distinguish changes in elemental composition, surface contamination, and can be combined with depth profiling. However, the technique is not widely accessible (i.e. costly equipment that requires supportive infrastructure), it also requires samples to be exposed to vacuum, which may limit its applicability. Quantitative analysis of XPS intensities has been demonstrated to provide understanding of film thicknesses and stoichiometries. Utilisation over extended periods can provided an extremely detailed interpretation of surface chemistry degradation [47].

2.3.2.2. Surface microstructure change

Changes to surface morphology can be investigated by the interpretation of 2D representations, or more accurately from quantitative 3D techniques. A simple approach to assessing changes to surface roughness is optical microscopy, however comprehensive assessment can only be made by using techniques such as confocal imaging as this provides a 3D representation [48]. Optical

Profilometry provides 3D topological information, able to facilitate analysis of surface damage, material displacement, in addition to other surface changes (e.g. self-healing materials). The technique is however limited to the analysis of relatively large surface features (i.e. >0.4 μm), due to diffraction limits of visible light [49]. 2D representations at higher resolution can be made using electron microscopy (**Figure 7**). Scanning electron microscopy (SEM) is very commonly used in the analysis of superhydrophobic morphology (**Figure 7**). Partial 3D interpretations can be attained through using tilted views of surfaces within SEM, however this does not provide reliable quantitative data [44, 50]. Contact profilometry can be used to provide 3D morphological data, and higher sensitivity analysis through atomic force microscopy (AFM—**Figure 7**) [51, 52]. These are similar techniques aimed at different morphological scales, with contact profilometry aimed at solely microscale roughness, and AFM able to measure nanoscale features.

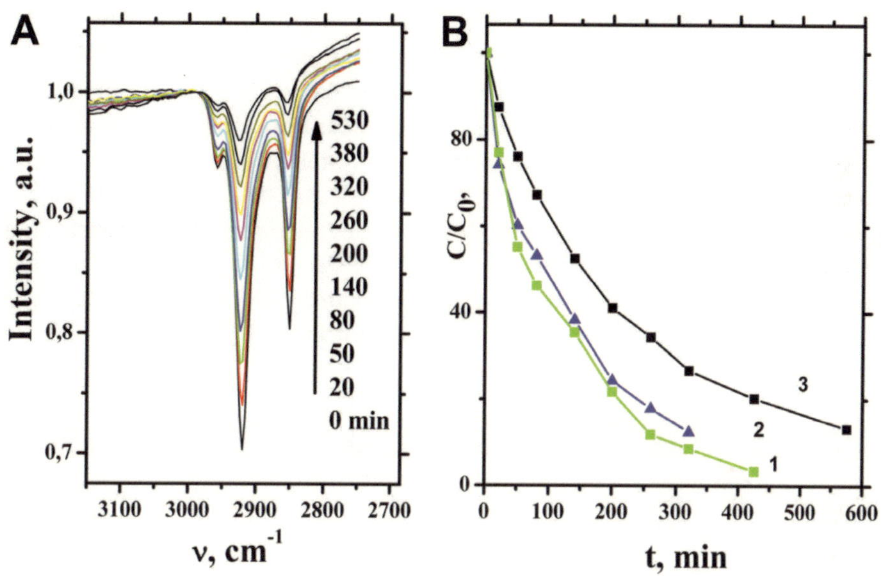

Figure 6. The degradation of surface bound molecules (stearic acid), monitored using infrared spectroscopy, on the surface of mesoporous TiO_2 films (A), and concentration analysis for steric acid degradation under UV light irradiation (B) on TiO_2 coatings formed under various conditions (1–3) [45].

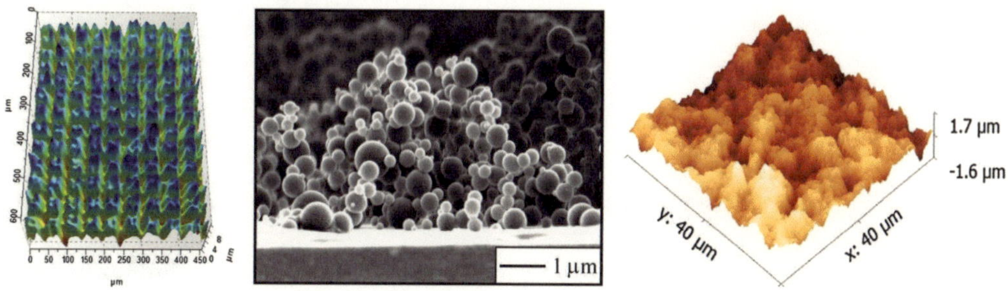

Figure 7. Surface change assessment techniques. Includes; confocal optical microscopy (left), side-on SEM imaging (centre), and AFM (right) [50, 54, 55].

These techniques (contact profilometry/AFM) lack the ability to resolve surface features it cannot make direct contact with, as they both use probes mounted perpendicular to the materials surface [53]. In addition, the ability to gain accurate/valuable information from the analysis depends on a range of factors, including; probe tip dimensions, mode of operation (contact, tapping, etc.), structural integrity of the surface material (potential movement/flex of analysis substrate), and contamination of probe tip by surface material, in addition to others [53].

2.3.2.3. In situ monitoring

Superhydrophobic degradation can be examined directly with WCA measurements, however, testing must be paused during this analysis (in addition to the using other analysis techniques mentioned in this Section 2.3.2) [56]. In situ monitoring of surface roughness is cannot be carried out easily, unless examining large surface features (e.g. using optical microscopy) [57]. Therefore, in situ monitoring is most valuable when examining reversible degradation. A commonly used example is the examination of surface reflectivity of a wetted surface [58]. The presence of air trapped at the water-surface interface notably produce mirror-like effects due to increased reflection of light from the water-air interface, when these materials are submerged in water (**Figure 8**). The monitoring the level of reflected light can be used to gauge the amount of trapped air, and its resultant variation. This testing has been implemented to show air layer stability in a range of conditions, including; submersion endurance, immersion depth testing, and flow measurements. The lifetime of the air layer at the surface upon submersion, has been shown to be highly dependent on both the test conditions (i.e. water flow, or applied pressure) and superhydrophobic surface morphology [59].

Figure 8. Analysis of recoverable (and non-recoverable) degradation *via* the use of surface reflectivity to monitor the surface plastron caused by superhydrophobic surfaces (A). The illustration (B) shows the distribution of micro/nanobubbles arranged across a superhydrophobic surface when submerged in water [59].

As mentioned in Section 2.3.1, water sheer can be used to induce superhydrophobic degradation [35]. Controlled water sheer (through the use of rheometry), has been used as a method for progressive monitoring of surface degradation under water sheer [34]. These techniques utilise the slip-length phenomenon demonstrated by superhydrophobic surfaces. Whereby, slip-length is gauged by monitoring the apparent viscosity of water on a superhydrophobic surface (measured by rheometry), which is lower than expected. The magnitude in the lowering of viscosity is directly proportional to the superhydrophobicity of the surface, and the amount of air trapped at the interface. As sheer stress is applied, both physical and reversible degradation occurs, which consequently reduces the measured slip-length. This measurement is carried out while the sheer stress testing is conducted.

3. Real world survivability

The potential application environments for superhydrophobic technologies are extremely diverse [1–3]. As a result, the requirements that ensure the long-term stability of these materials can be focus on a range of different priorities. This section highlights potential application environments, the associated expected tolerances, and the introduction of the associated engineering challenges.

3.1. Application environments

The potential applications areas of superhydrophobic materials, and the expected operational conditions, stem from their numerous functional properties (self-cleaning, antibiofouling, drag-reduction, oil-water separation, etc.) [1–3]. These areas can be classified with respect to environmental tolerances; (i) high (broad-spectrum), (ii) low, and (iii) high (application specific challenges). The environmental stimuli able to influence surface degradation include; fluid flow (water/air), physical (solid) contact, chemical/biological exposure, and environmental changes (pressure, temperature, etc.). These expected tolerance classifications are briefly summarised below.

3.1.1. Broad-spectrum high tolerance environments

This covers a wide range of potential applications, but includes those that require; manual handling, abrasive interaction, sheering water flow, and multi-environment resilience [60]. Currently, no commercially available superhydrophobic products exhibit a generally high environmental tolerance. An example application would be the superhydrophobic treatment of textiles, which requires resilience to handling, abrasion, varied environmental exposure, etc. [61]. Additionally, for superhydrophobic textiles, the loss of functionality when laundered (i.e. water sheer and chemical exposure) would also be a concern [62]. As a result, currently all commercial superhydrophobic fabric treatments are marketed as temporary, requiring reapplication after prolonged wearing, or after laundering [63]. The use of superhydrophobic surfaces in extreme environments would prove hugely beneficial. Aerospace surfaces are an example of a highly challenging, yet highly relevant operational environment, this includes;

high fluid sheer, rapid temperature change, and airborne particle impact [64, 65]. Affording self-cleaning, anti-fogging, and anti-icing properties in these applications would be highly desired, however this currently exceeds technological capabilities.

3.1.2. Low tolerance environments

The current standard of commercial superhydrophobic technology already enables application in low tolerance areas [63]. As described above, the temporary treatment of textiles can be readily achieved. However, superhydrophobicity cannot be maintained by these treatments under garment tolerances over long periods. These types of coating, as well as many other surface treatments reported in the literature, have been shown to resist low intensity stimuli (e.g. water flow, or adhesive testing) [63, 66]. Surfaces that do not experience physical interactions (surface-solid contact) would be an area for potential application. An example would include the internal surface of liquid carrying pipes (e.g. water drainage pipes), omitting any high sheer forces, and any application specific challenges (described below) [67].

3.1.3. Application specific challenges

Surface degradation *via* physical action, or fluid shear, are commonly considered as target forces to resist [27, 28, 34]. However, there are a range of degradation sources specific to particular applications. Surface chemistry variation can be accompanied by tremendous changes in surface wettability. This commonly occurs *via* two main mechanisms; (i) surface reactions to remove and/or alter hydrophobic chemistry, or (ii) the addition of surface material to mask effective surface chemistry [18, 19]. The former can occur upon exposure to harsh chemicals or other degradation route (e.g. highly basic conditions, or photocatalytic degradation), but can also degrade readily under ambient conditions over time [9, 44, 68]. Therefore, a thorough consideration of superhydrophobic resilience would include the stability of surface functionalisation, the compatibility between the surface coating and underlying material, in addition to the effect of any reactive species present within the local environment. Surface fouling is often accompanied by an associated reduction in surface hydrophobicity [18]. Biofouling in particular is a rapidly escalating process, whereby any initial fouling can encourage further biological surface attachment. Fouling prevention has been shown possible with respect to specific surface contamination sources. For example, there are many literature examples of the removal of particulates, or dyes from the surface [69]. However, the design of general fouling prevention is not facile, as it depends on a multitude of factors (surface morphology/chemistry, contamination source, and contaminant delivery method). This is evident when considering the biofouling process, which is a multifaceted mechanism (**Figure 9**) [70]. The primary fouling prevention target are small molecules that are able to contaminate the surface, conditioning it for microbial attachment. Antifouling in subsequent biofouling focuses on preventing the attachment of microbial species.

3.2. Expected tolerances

The expected resilience of a surface can be considered in three categories; (i) physical degradation, (ii) reversible degradation, and (iii) chemical degradation.

http://dx.doi.org/10.5772/intechopen.80746

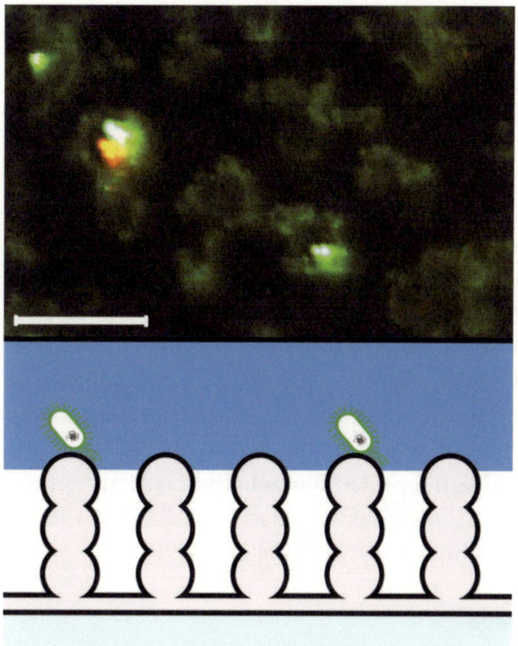

Figure 9. The attachment of bacteria to superhydrophobic surfaces is shown to preferentially occur at the tops of surface protrusions. This is caused by superhydrophobic wetting mechanisms, which leaves these areas as the only viable attachment points. This has been demonstrated in bacterial attachment assays (upper), and shown by the illustration (lower) [71].

Physical degradation can occur *via* a range of mechanisms (Section 2.3.1), and the challenge of optimising physical resilience is a multifaceted consideration. However, there are some key principles in surface engineering that can be targeted. The morphology of superhydrophobic surfaces can be imagined as an array of surface protrusions [10]. Physical degradation of these features can occur *via* compression/destruction, or removal, which is dependent on both the structural integrity of individual surface features, and their attachment to the underlying substrate [10, 11]. The real-world applicability of superhydrophobic materials requires the survival of nano/microstructured features. A key consideration for surface design is the exact forces associated with a particular application. Furthermore, considering whether a given design route can offer the required resilience. The surface pressures applied during heavy-duty manual handling processes can range from 100 to 400 kPa [72]. The pressure applied to individual surface features is expected to be higher than this, due to an uneven morphology caused by surface roughness.

The conditions required for reversible degradation can be predicted for highly ordered surface structures, as the hydrodynamic factors that dictate air entrapment at a surface are well established [73, 74]. Most materials aimed commercialisation are made up of randomised surface features, and provide an array of trapped air environments. Therefore, practical testing of these surfaces is the only reliable method for assessing trapped air stability [31]. Generally, critical pressure is used to classify the hydrodynamic pressure required to remove trapped air. Critical pressure is the pressure required for conversion of Cassie-Baxter to Wenzel wetting. For superhydrophobic materials this critical pressure is generally greater than 0.5 kPa, however nanosized features can increase this to ~300 kPa and greater [75]. A reason why superhydrophobic leaves may also incorporate nano-roughness in their surface morphologies, is that this allows protection from impact of rain droplets, where the applied pressure range is 10–1000 kPa [75].

Stable surface chemistry can be ensured *via* careful consideration of the hydrophobic surface treatments [16, 17]. Facilitated through the incorporation of strong chemical bonding between the surface coatings and underlying morphology, or the exploitation of strong adhesive forces. There are many examples in the literature that report resilience to a range of chemical environments (e.g. extreme pH or solvent exposure) [76]. As mentioned in Section 3.1, a general approach for chemical/biological fouling prevention is not straightforward. Successful fouling prevention can be carried out if specific contaminants are targeted. Many literature examples examine the removal of particulates that are loosely arranged on the surfaces [69]. In addition, the prevention of bacterial attachment has also been studied, however no comprehensive approach has been reported that relies upon solely superhydrophobic antifouling properties [77]. Many superhydrophobic surfaces with high antifouling properties also incorporate biocidal species within the reported materials [78].

3.3. Engineering challenge

The requirements for truly resilient superhydrophobic surfaces that are able to function in real-world applications are extremely demanding, particularly if a general high tolerance is required. Numerous approaches for resilience have been developed (Section 4), however a multi-faceted research effort is required to probe the extremities of physical possibility with respect to maximising resilience. This investigation incorporates aspects of chemistry, materials science, materials engineering, fluid mechanics, and microbiology, in addition to other relevant areas. The development of real-world materials requires a drive toward the standardised testing and reporting of resilience, this allows for formulation of an effective design approach.

4. Approaches for robustness

4.1. Hardening materials

The structural integrity of superhydrophobic surface morphology is key to the magnitude of their physical resilience [10]. An improvement to the integrity of a structured surface can be made by modifying the composition of the component materials. This has been achieved through using inherently high strength materials to construct a surface (Section 4.1.1), or *via* the utilisation of strengthening modifications (Section 4.1.2—**Figure 10**).

4.1.1. High strength materials

A range of factors determines the strength of a material (e.g. chemical bonds/structure, intra/intermolecular binding, etc.), this directly affects the materials stress tolerances [82, 83]. As a result, there is a range of candidate materials for engineering resilient superhydrophobic materials. Materials formed by strong compression of superhydrophobic components provide an

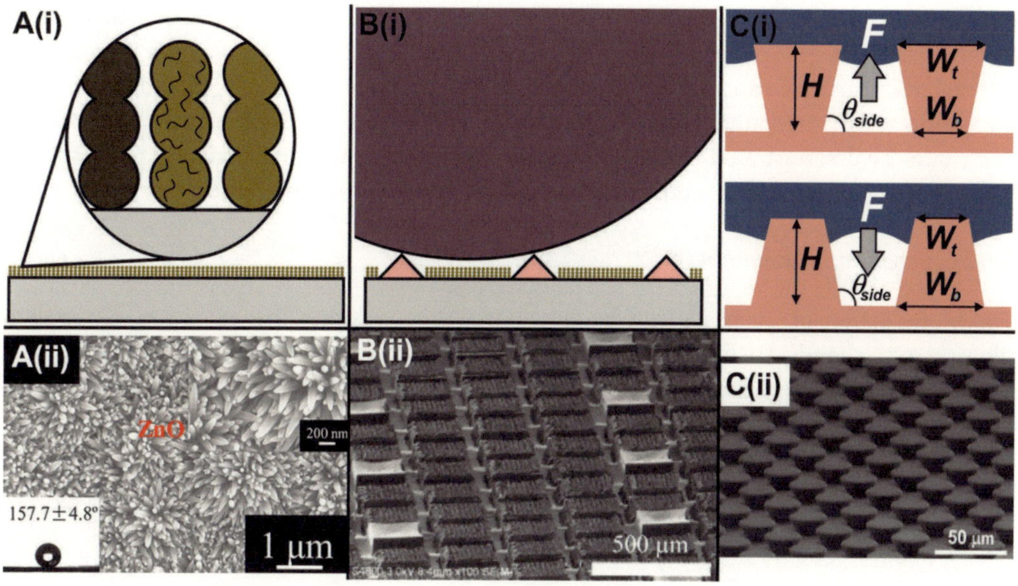

Figure 10. Approaches for the formulation of robust superhydrophobic materials. Includes; (A) hardening materials (inherent strength, or via physical/chemical strengthening additives), (B) protective surface features, and (C) the strategic design of optimised surface features [79–81].

assembly method, able to greatly increase their physical resilience. The compression of hydrophobic polymer nanoparticle mixtures has been used to form superhydrophobic disks [84]. These are reported to have high robustness, able to withstand cutting, abrasion, and chemical exposure. The materials are compacted, a high surface roughness remains, and maintain an inherently hydrophobic surface chemistry, facilitating superhydrophobicity. The resistance to physical degradation is explained as a result of the fabrication technique, which relies upon pressing pressures of up to 40 MPa, and the highly condensed material that results. However, within this example, the resilience testing carried out does not provide quantifiable resilience data [84].

Electroplating has also been used to fabricate highly robust coatings on to conductive substrates. The electroplating techniques can provide strong underlying connection to the substrate [85, 86]. Highly rough surface features have been reported *via* numerous literature electroplating routes, with superhydrophobicity resulting from the addition of a hydrophobic surface coating [87]. Superhydrophobic surfaces fabricated *via* electrodeposition exhibit a range of surface features, including; nanopillars, nanoflakes, and various crystallite formations (**Figure 10**). An example reporting high resilience utilises zinc/nickel/cobalt composites to generate superhydrophobic materials able to resist prolonged surface abrasion. The physical robustness is explained through the formation of a highly strong composite material, in combination with effective substrate binding [79].

Another example reported to enhance both physical and chemical resilience, is the utilisation of protective surface treatments. Silanisation has been effectively applied to melamine sponges [88]. Whereby the sponge microstructure acts as a roughening template, with hydrophobic silanisation [$-(Si(R)-O)_n-$, where $R = -(CH_2)_{17}CH_3$] delivering superhydrophobicity. The sponge is demonstrated as a selective oil absorbent material, with an affinity for a range of hydrophobic liquids (includes; toluene, chloroform, diesel, and motor oil). Upon absorption of the liquid, it is subsequently removed by applying a compressive force; this deforms the internal microstructure and is a potential cause of degradation. The sponges undergo successive liquid absorption/removal cycles (up to 1000 iterations) without the loss of functionality. The resilience demonstrated by the sponges is explained through the flexibility properties of the melamine sponges, and robust chemistry of the surface protecting chemistry [88].

4.1.2. Strengthening modifications

The modification of materials can be classified by considering the alteration as either a physical, or a chemical, variation to an existing material. In a physical addition, the chemical structure of the original components does not change (e.g. incorporation of carbon nanotubes into polymer matrix) [89]. Chemical additions are therefore associated with a change in the chemical structure of one or more original components (e.g. the doping of semi-conductor structure) [90].

4.1.2.1. Physical addition

The addition of strengthening components into an existing structure operate *via* a combination of two main mechanisms. The components can fortify the superhydrophobic microstructure by cohesive binding, in addition to the impregnated material imparting intrinsic resilience [82, 83]. The exact balance of these factors varies depending on the particular materials combination. A commonly used physical strengthening additive are carbon nanotubes, as they have an intrinsically high structural integrity, and can also contribute additional roughness to the surface microstructure. The incorporation of carbon nanotubes has been shown to increase resilience to applied forces, while improving surface hydrophobicity [89, 91].

4.1.2.2. Chemical addition

The microstructure and resultant resilience of a superhydrophobic material can be directly determined by the chemical structure of the composite materials. Chemically altered components have been utilised throughout the literature as a means to increase surface hydrophobicity, and improved resilience [90]. Porous organic polymers (POPs) are an example of the importance of chemical structure, as some POPs are susceptible to hydrolytic degradation [92]. The protection of these porous polymers can be achieved through the incorporation of hydrophobic species into the polymer, which reduce the chemical degradation—achieving water contact angles of 152°. This is demonstrated to greatly extend the applicability of the reported POP material [92].

4.2. Structuring materials

The architecture of surface morphology can play a significant role in determining a materials resilience (**Figure 10**). This has led to the development of a range of surface design approaches to secure the physical robustness of superhydrophobic materials.

4.2.1. *Protective surface features*

The dual-scale roughness of many superhydrophobic surfaces can be exploited to generate bulkier protective features (**Figure 10**) [93, 80]. The principle behind this approach is to utilise larger, more resilience morphological features to act as the sacrificial points of contact when handled manually. Two types of surface architecture that utilise this approach include sacrificial points of contact that emanate from the surfaces, and surfaces with inaccessible superhydrophobic divots. The superhydrophobic divot approach has been accomplished *via* laser etching of metals followed by hydrophobic surface treatment. Whereby the non-etched areas act as protective contact points [93]. Eminent structuring has been achieved through a number of fabrication techniques, this includes micromachining, followed by moulding of polypropylene (**Figure 10**). The result was a multi-scale surface roughness generating samples with large features able withstand mechanical compression up to 20 MPa, and abrasive wear tests up to 120 kPa [80].

4.2.2. *Optimised surface architectures*

The surface microstructure has been shown to play a significant role in increasing the physical strength of surface features. The degradation of surface microstructures can occur in a range of ways (i.e. shearing, mushrooming, and splitting), with the surface architecture influencing the route of degradation [94]. In the fabrication of thin films and coatings, the strength of adhesion to the substrate must also be considered. Many high strength coatings are supported by the inclusion of binding layers to improve adhesion [95, 96]. Many superhydrophobic coatings from the literature incorporate surface treatments to improve substrate binding, this includes; molecular layers (e.g. SAMs), polymer films, and commercial adhesives [56, 97, 98].

Strategically designed surface architectures are highly challenging to formulate, particularly when designing intensely water repelling, as they require dual-scale roughness (i.e. micro/nanoroughness) [1–3]. Examples of utilising exclusively micrometre-sized structures to ensure robustness have been reported [99]. One approach implements microfabrication to generate a polydimethylsiloxane pillar array (**Figure 10**), further optimised by the addition of a fluoroalkyl film. Whereby substantial micropillars, in combination with targeted geometry prevent physical degradation, and the loss of trapped air [81]. An example that targets both strong substrate binding, and robust surface features, utilises a multi-step coating deposition [100]. The technique ensures strong substrate adhesion (using a solution growth of silica onto glass), whereby surface roughness is generated by depositing three forms of silica (mesoporous thin film, nanoparticle dispersion, and mesoporous nanosheet dispersion), which is then locked into place by a final coating of silica. Hydrophobic surface treatment is then used to generate the final superhydrophobic material, able to with stand abrasive forces [100].

4.2.3. *Preventing air loss*

The stability of air trapped at a wetted superhydrophobic interface can be assessed experimentally (Section 2.3.2), however cannot be fully predicted, especially if morphological features are distributed arbitrarily [23]. Generally, it is observed that statistically larger volumes of trapped air are more easily removed then smaller volumes [34]. However, a combination of highly water repellent surface chemistries, and strategically designed morphological features, can be used to prevent air loss [81]. High aspect ratio needle structure (nanoneedles) with hydrophobic surface chemistries have been shown to provide a high stability for trapped air, through a combination of intrinsically high WCAs, and a wide range of trapped air environments [101]. Superhydrophobic surfaces fabricated from arrays of holes have also been demonstrated to provide a high stability of trapped air. This approach is able to provide a high proportion of trapped air that is firmly positioned, requiring a high hydrostatic pressure for its removal [102].

4.3. Responsive surface morphologies

This section summarises materials that exhibit and engineered response to physical stimuli, preserving high surface roughness, and/or hydrophobic surface chemistry, thus maintaining superhydrophobicity.

4.3.1. Mobile surface features

High intensity physical abrasion applied to superhydrophobic materials will tend to result in permanent degradation in wetting properties, due to damage of the surface features [26–28]. An approach reported in the literature utilises surface features that are able to be displaced from one position to another on the surface. Where roughening components are repositioned, the degree of the surface roughness is maintained—the result is a transient microstructure (**Figure 11**) [56]. The utilisation of nanomaterials in the fabrication of superhydrophobic materials is widespread throughout the literature [103]. A coating assembled from solely hydrophobic nanoparticles (no binders/adhesives), has an extremely low resilience, and could be completely removed if manually handled [104]. This has led to the inclusion of adhesives and binders to hold the nanoparticle coating to the substrate. Resilience testing has reviled that the adhesive layer also prevents displaced particles from leaving the surface, with the rearrangement causing no change to observed WCAs. Loss of surface material over extensive abrasive repetitions is reported [105].

4.3.2. Regenerative structures/chemistry

The physical/chemical degradation of superhydrophobic materials and the subsequent reduction in surface hydrophobicity, is likely to be a permanent process [10, 46]. This is a result of a loss in surface roughness, and/or a transformation to lower hydrophobicity surface chemistry. Recovery from inflicted damage can be engineered into superhydrophobic materials by the incorporation of regenerative pathways. Self-healing materials have been reported *via* the inclusion of hydrophobic resins within the surface microstructure [107–110]. The leeching of these resins when surface damage occurs shields the exposed area. Although this leeching approach has demonstrated success, many routes do not involve the recovery of surface roughness. As a result, a reduction in WCAs are generally observed when extended abrasion testing is carried out [107–110]. The comprehensive recovery of surface hydrophobicity requires the regeneration of surface roughness. An example that targets this utilises multicomponent mixtures (including; hydrophobic elastomer, and polyhedral oligomeric

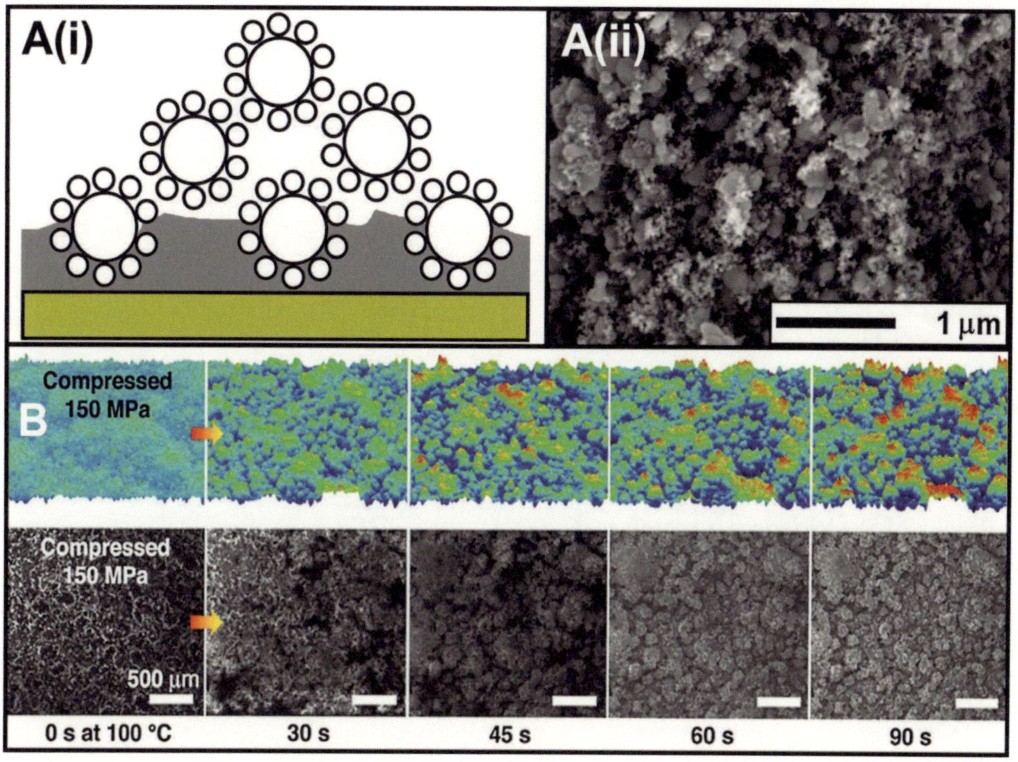

Figure 11. Surface morphologies engineered to provide reactive resilience, through the use of (A) transient micro/nano-structures, and (B) regenerative surfaces [56, 106].

silsesquioxane) that are able to regenerate surface roughness even when the material is fully abraded. The regeneration is facilitated through strategic selection of the mixture components, and an optimised regeneration protocol (**Figure 11**) [106].

5. Conclusions

This chapter has provided a summary of approaches for evaluating and engineering resilience within superhydrophobic materials. The development of superhydrophobic surfaces robust enough to withstand even light manual handling has the potential to tremendously extend their current applicability. There are huge amounts of literature examples asserting varying degrees of resilience, many utilising arbitrary testing methods that provide minimal quantitative resilience information (Section 2.3.1). Although the engineering of superhydrophobic surfaces has made progress toward materials useable under real-world conditions, this advancement would be accelerated by the adoption of standardised resilience testing techniques.

The approaches to assuring resilience to physical/chemical/reversible degradation are well established (Section 4). The development of these principles (i.e. improving; strength, versatility, air-layer stability, etc.) is a tremendous focus for researchers in this area. Superhydrophobic materials have been shown to have extraordinary potential in many applications (e.g. antibiofouling, self-cleaning, drag-reduction, etc.). Exploring the limitations of micro/nanostructure resilience will be critical in determining the ultimate applicability of superhydrophobic technology.

Acknowledgements

CRC would like to acknowledge The Royal Society (RG160758), and EPSRC for funding.

Conflict of interest

CRC would like to declare no conflicts of interest.

Author details

Colin R. Crick

Address all correspondence to: c.crick@liverpool.ac.uk

Materials Innovation Factory, University of Liverpool, Liverpool, United Kingdom

References

[1] Simpson JT, Hunter SR, Aytug T. Superhydrophobic materials and coatings: A review. Reports on Progress in Physics. 2015;**78**(8):86501

[2] Su B, Tian Y, Jiang L. Bioinspired interfaces with superwettability: From materials to chemistry. Journal of the American Chemical Society. 2016;**138**(6):1727-1748

[3] Liu M, Wang S, Jiang L. Nature-inspired superwettability systems. Nature Reviews Materials. 2017;**2**(7):17036

[4] Carrascosa LAM, Facio DS, Mosquera MJ. Producing superhydrophobic roof tiles. Nanotechnology. 2016;**27**(9):95604

[5] Samaha MA, Tafreshi HV, Gad-el-Hak M. Superhydrophobic surfaces: From the lotus leaf to the submarine. Comptes Rendus Mécanique. 2012;**340**(1):18-34

[6] Crick CR, Parkin IP. Preparation and characterisation of super-hydrophobic surfaces. Chemistry—A European Journal;**16**(12):3568-3588

[7] Latthe SS, Terashima C, Nakata K, Fujishima A. Superhydrophobic surfaces developed by mimicking hierarchical surface morphology of lotus leaf. Molecules. 2014;**19**(4):4256-4483

[8] Gao X, Xu L-P, Xue Z, Feng L, Peng J, Wen Y, et al. Dual-scaled porous nitrocellulose membranes with underwater superoleophobicity for highly efficient oil/water separation. Advanced Materials. 2014;**26**(11):1771-1775

[9] Davis A, Yeong YH, Steele A, Loth E, Bayer IS. Spray impact resistance of a superhydrophobic nanocomposite coating. AICHE Journal. 2014;**60**(8):3025-3032

[10] Quéré D. Wetting and roughness. Annual Review of Materials Research. 2008;**38**(1):71-99

[11] Chu Z, Seeger S. Superamphiphobic surfaces. Chemical Society Reviews. 2014;**43**(8): 2784-2798

[12] Makkonen L. Young's equation revisited. Journal of Physics. Condensed Matter. 2016; **28**(13):135001

[13] Wenzel RN. Resistance of solid surfaces to wetting by water. Industrial and Engineering Chemistry. 1936;**28**(8):988-994

[14] Cassie ABD, Baxter S. Wettability of porous surfaces. Transactions of the Faraday Society. 1944;**40**(0):546-551

[15] Aoki T, Okumura Y, Shioi M, Kawamura T. Evaluation of PEG-SAM stabilities and inhibition of nonspecific protein adsorption during storage for 2 months. Chemistry Letters. 2015;**44**(12):1661-1663

[16] Vemuri S, Kim KJ, Wood BD, Govindaraju S, Bell TW. Long term testing for dropwise condensation using self-assembled monolayer coatings of n-octadecyl mercaptan. Applied Thermal Engineering. 2006;**26**(4):421-429

[17] Jafari R, Momen G, Farzaneh M. Durability enhancement of icephobic fluoropolymer film. Journal of Coating Technology and Research. 2016;**13**(3):405-412

[18] Ista LK, Callow ME, Finlay JA, Coleman SE, Nolasco AC, Simons RH, et al. Effect of substratum surface chemistry and surface energy on attachment of marine bacteria and algal spores. Applied and Environmental Microbiology. 2004;**70**(7):4151-4157

[19] Gryta M. Effect of iron oxides scaling on the MD process performance. Desalination. 2007;**216**(1):88-102

[20] Qian H, Zhu Y, Wang H, Song H, Wang C, Liu Z, et al. Preparation and antiscaling performance of superhydrophobic poly(phenylene sulfide)/polytetrafluoroethylene composite coating. Industrial and Engineering Chemistry Research. 2017;**56**(44):12663-12671

[21] Giacomello A, Schimmele L, Dietrich S, Tasinkevych M. Perpetual superhydrophobicity. Soft Matter. 2016;**12**(43):8927-8934

[22] Vrancken RJ, Kusumaatmaja H, Hermans K, Prenen AM, Pierre-Louis O, Bastiaansen CWM, et al. Fully reversible transition from Wenzel to Cassie–Baxter states on corrugated superhydrophobic surfaces. Langmuir. 2010;**26**(5):3335-3341

[23] Bormashenko E. Progress in understanding wetting transitions on rough surfaces. Advances in Colloid and Interface Science. 2015;**222**:92-103

[24] Mayser MJ, Barthlott W. Layers of air in the water beneath the floating fern salvinia are exposed to fluctuations in pressure. Integrative and Comparative Biology. 2014; **54**(6):1001-1007

[25] Xiang Y, Huang S, Lv P, Xue Y, Su Q, Duan H. Ultimate stable underwater superhydrophobic state. Physical Review Letters. 2017;**119**(13). Article ID: 036101. Available from: https://link.aps.org/doi/10.1103/PhysRevLett.119.134501

[26] Mortazavi V, Khonsari MM. On the degradation of superhydrophobic surfaces: A review. Wear. 2017;**372-373**:145-157

[27] Wang N, Lu Y, Xiong D, Carmalt CJ, Parkin IP. Designing durable and flexible superhydrophobic coatings and its application in oil purification. Journal of Materials Chemistry A. 2016;**4**(11):4107-4116

[28] Steele A, Nayak BK, Davis A, Gupta MC, Loth E. Linear abrasion of a titanium superhydrophobic surface prepared by ultrafast laser microtexturing. Journal of Micromechanics and Microengineering. 2013;**23**(11):115012

[29] Morgeneyer M, Shandilya N, Chen Y-M, Le Bihan O. Use of a modified taber abrasion apparatus for investigating the complete stress state during abrasion and in-process wear particle aerosol generation. Chemical Engineering Research and Design. 2015;**93**:251-256

[30] Lei W, Jia Z-H, He J-C, Cai T-M, Wang G. Vibration-induced Wenzel-Cassie wetting transition on microstructured hydrophobic surfaces. Applied Physics Letters. 2014; **104**(18):181601

[31] Li J. Characterization for Cassie-Wenzel wetting transition based on the force response in the process of squeezing liquid drops by two parallel superhydrophobic surfaces. The Review of Scientific Instruments. 2016;**87**(6):65108

[32] Crick CR, Parkin IP. Water droplet bouncing—A definition for superhydrophobic surfaces. Chemical Communications. 2011;**47**(44):12059-12061

[33] Crick CR, Parkin IP. Relationship between surface hydrophobicity and water bounces—A dynamic method for accessing surface hydrophobicity. Journal of Materials Chemistry A. 2013;**1**(3):799-804

[34] Xu HR, Crick CJ, Poole R. Evaluating the resilience of superhydrophobic materials using the slip-length concept. Journal of Materials Chemistry A. 2018;**6**(10):4458-4465

[35] Sakai M, Nakajima A, Fujishima A. Removing an air layer from a superhydrophobic surface in flowing water. Chemistry Letters. 2010;**39**(5):482-484

[36] Antonini C, Villa F, Marengo M. Oblique impacts of water drops onto hydrophobic and superhydrophobic surfaces: Outcomes, timing, and rebound maps. Experiments in Fluids. 2014;**55**(4):1713

[37] Shirtcliffe NJ, McHale G, Newton MI, Zhang Y. Superhydrophobic copper tubes with possible flow enhancement and drag reduction. ACS Applied Materials & Interfaces. 2009;**1**(6):1316-1323

[38] Janjua ZA, Turnbull B, Choy K-L, Pandis C, Liu J, Hou X, et al. Performance and durability tests of smart icephobic coatings to reduce ice adhesion. Applied Surface Science. 2017;**407**:555-564

[39] Gemici Z, Shimomura H, Cohen RE, Rubner MF. Hydrothermal treatment of nanoparticle thin films for enhanced mechanical durability. Langmuir. 2008;**24**(5):2168-2177

[40] Li Y, Li Q, Zhang C, Cai P, Bai N, Xu X. Intelligent self-healing superhydrophobic modification of cotton fabrics via surface-initiated ARGET ATRP of styrene. Chemical Engineering Journal. 2017;**323**:134-142

[41] Babu DJ, Mail M, Barthlott W, Schneider JJ. Superhydrophobic vertically aligned carbon nanotubes for biomimetic air retention under water (Salvinia effect). Advanced Materials Interfaces. 2017;**4**(13):1700273

[42] Zeng X, Xu G, Gao Y, An Y. Surface wettability of (3-aminopropyl)triethoxysilane self-assembled monolayers. The Journal of Physical Chemistry. B. 2011;**115**(3):450-454

[43] Devaprakasam D, Sampath S, Biswas SK. Thermal stability of perfluoroalkyl silane self-assembled on a polycrystalline aluminum surface. Langmuir. 2004;**20**(4):1329-1334

[44] Crick CR, Bear JC, Kafizas A, Parkin IP. Superhydrophobic photocatalytic surfaces through direct incorporation of titania nanoparticles into a polymer matrix by aerosol assisted chemical vapor deposition. Advanced Materials. 2012;**24**(26):3505-3508

[45] Smirnova N, Fesenko T, Zhukovsky M, Goworek J, Eremenko A. Photodegradation of stearic acid adsorbed on superhydrophilic TiO_2 surface: In situ FT-IR and LDI study. Nanoscale Research Letters. 2015;**10**(1):500

[46] Xu W, Chen X, Cai S, Chen J, Xu Z, Jia H, et al. Superhydrophobic titania nanoparticles for fabrication of paper-based analytical devices: An example of heavy metals assays. Talanta. 2018;**181**:333-339

[47] Petrovykh DY, Kimura-Suda H, Opdahl A, Richter LJ, Tarlov MJ, Whitman LJ. Alkanethiols on platinum: Multicomponent self-assembled monolayers. Langmuir. 2006;**22**(6): 2578-2587

[48] Duparré A, Ferre-Borrull J, Gliech S, Notni G, Steinert J, Bennett JM. Surface characterization techniques for determining the root-mean-square roughness and power spectral densities of optical components. Applied Optics. 2002;**41**(1):154-171

[49] Saurín N, Sanes J, Bermúdez MD. Self-healing of abrasion damage in epoxy resin–ionic liquid nanocomposites. Tribology Letters. 2015;**58**(1):4

[50] Crick CR, Parkin IP. Aerosol assisted deposition of melamine-formaldehyde resin: Hydrophobic thin films from a hydrophilic material. Thin Solid Films. 2011;**519**(7): 2181-2186

[51] Ponte DC, Meyer DML. Frictional behavior and topography of porous polyurethane on copper and silicon dioxide articulating contacts. Journal of Tribology. 2016;**138**(3): 31604-0-31604-9

[52] Hamers RJ. Scanned probe microscopies in chemistry. The Journal of Physical Chemistry. 1996;**100**(31):13103-13120

[53] Maver U, Velnar T, Gaberšček M, Planinšek O, Finšgar M. Recent progressive use of atomic force microscopy in biomedical applications. TrAC Trends in Analytical Chemistry. 2016;**80**:96-111

[54] Ta VD, Dunn A, Wasley TJ, Li J, Kay RW, Stringer J, et al. Laser textured superhydrophobic surfaces and their applications for homogeneous spot deposition. Applied Surface Science. 2016;**365**:153-159

[55] Li J, Wei Y, Huang Z, Wang F, Yan X, Wu Z. Electrohydrodynamic behavior of water droplets on a horizontal super hydrophobic surface and its self-cleaning application. Applied Surface Science. 2017;**403**:133-140

[56] Lu Y, Sathasivam S, Song J, Crick CR, Carmalt CJ, Parkin IP. Robust self-cleaning surfaces that function when exposed to either air or oil. Science. 2015;**347**(6226):1132-1135

[57] Sabbah A, Youssef A, Damman P. Superhydrophobic surfaces created by elastic instability of PDMS. Applied Sciences. 2016;**6**(5):152

[58] Hokmabad BV, Ghaemi S. Effect of flow and particle-plastron collision on the longevity of superhydrophobicity. Scientific Reports. 2017;**7**:41448

[59] Poetes R, Holtzmann K, Franze K, Steiner U. Metastable underwater superhydrophobicity. Physical Review Letters. 2010;**105**(16):166104

[60] Tian X, Verho T, Ras RHA. Moving superhydrophobic surfaces toward real-world applications. Science. 2016;**352**(6282):142-143

[61] Zhou X, Zhang Z, Xu X, Guo F, Zhu X, Men X, et al. Robust and durable superhydrophobic cotton fabrics for oil/water separation. ACS Applied Materials & Interfaces. 2013;**5**(15):7208-7214

[62] Zeng C, Wang H, Zhou H, Lin T. Self-cleaning, superhydrophobic cotton fabrics with excellent washing durability, solvent resistance and chemical stability prepared from an SU-8 derived surface coating. RSC Advances. 2015;**5**(75):61044-61050

[63] Gupta R, Vaikuntanathan V, Sivakumar D. Superhydrophobic qualities of an aluminum surface coated with hydrophobic solution NeverWet. Colloids and Surfaces A: Physicoc hemical and Engineering Aspect. 2016;**500**:45-53

[64] Zhao L, Liu WL, Zhang LD, Yao JS, Xu WH, Wang XQ, et al. Fabrication of superhydrophobic and conductive surface based on carbon nanotubes. Colloids and Surfaces A: Ph ysicochemical and Engineering Aspect. 2013;**423**:69-76

[65] Larson C, Smith JR, Armstrong GJ. Current research on surface finishing and coatings for aerospace bodies and structures—A review. Transactions of the IMF. 2013;**91**(3):120-132

[66] Esmeryan KD, Castano CE, Bressler AH, Abolghasemibizaki M, Mohammadi R. Rapid synthesis of inherently robust and stable superhydrophobic carbon soot coatings. Applied Surface Science. 2016;**369**:341-347

[67] Geraldi NR, Dodd LE, Xu BB, Wells GG, Wood D, Newton MI, et al. Drag reduction properties of superhydrophobic mesh pipes. Surface Topography: Metrology and Properties. 2017;**5**(3):34001

[68] Ma J, Zhang XY, Wang DP, Zhao DQ, Ding DW, Liu K, et al. Superhydrophobic metallic glass surface with superior mechanical stability and corrosion resistance. Applied Physics Letters. 2014;**104**(17):173701

[69] Gao S, Huang J, Li S, Liu H, Li F, Li Y, et al. Facile construction of robust fluorine-free superhydrophobic TiO_2@fabrics with excellent anti-fouling, water-oil separation and UV-protective properties. Materials and Design. 2017;**128**:1-8

[70] Römling U, Balsalobre C. Biofilm infections, their resilience to therapy and innovative treatment strategies. Journal of Internal Medicine. 2012;**272**(6):541-561

[71] Crick CR, Ismail S, Pratten J, Parkin IP. An investigation into bacterial attachment to an elastomeric superhydrophobic surface prepared via aerosol assisted deposition. Thin Solid Films. 2011;**519**(11):3722-3727

[72] Fransson-Hall C, Kilbom Å. Sensitivity of the hand to surface pressure. Applied Ergonomics. 1993;**24**(3):181-189

[73] Lopes DM, Ramos SMM, de Oliveira LR, Mombach JCM. Cassie–Baxter to Wenzel state wetting transition: A 2D numerical simulation. RSC Advances. 2013;**3**(46):24530-24534

[74] Kwon H-M, Paxson AT, Varanasi KK, Patankar NA. Rapid deceleration-driven wetting transition during pendant drop deposition on superhydrophobic surfaces. Physical Review Letters. 2011;**106**(3):36102

[75] Zheng Q-S, Yu Y, Zhao Z-H. Effects of hydraulic pressure on the stability and transition of wetting modes of superhydrophobic surfaces. Langmuir. 2005;**21**(26):12207-12212

[76] Guo L, Yuan W, Li J, Zhang Z, Xie Z. Stable superhydrophobic surfaces over a wide pH range. Applied Surface Science. 2008;**254**(7):2158-2161

[77] Hizal F, Rungraeng N, Lee J, Jun S, Busscher HJ, van der Mei HC, et al. Nanoengineered superhydrophobic surfaces of aluminum with extremely low bacterial adhesivity. ACS Applied Materials & Interfaces. 2017;**9**(13):12118-12129

[78] Ozkan E, Crick CC, Taylor A, Allan E, Parkin IP. Copper-based water repellent and antibacterial coatings by aerosol assisted chemical vapour deposition. Chemical Science. 2016;**7**(8):5126-3131

[79] Xiang T, Han Y, Guo Z, Wang R, Zheng S, Li S, et al. Fabrication of inherent anticorrosion superhydrophobic surfaces on metals. ACS Sustainable Chemistry & Engineering. 2018;**6**(4):5598-5606

[80] Huovinen E, Takkunen L, Korpela T, Suvanto M, Pakkanen TT, Pakkanen TA. Mechanically robust superhydrophobic polymer surfaces based on protective micropillars. Langmuir. 2014;**30**(5):1435-1443

[81] Im M, Im H, Lee J-H, Yoon J-B, Choi Y-K. A robust superhydrophobic and superoleophobic surface with inverse-trapezoidal microstructures on a large transparent flexible substrate. Soft Matter. 2010;**6**(7):1401-1404

[82] Meyers MA, Mishra A, Benson DJ. Mechanical properties of nanocrystalline materials. Progress in Materials Science. 2006;**51**(4):427-556

[83] Zhao J, Wang F, Zhang X, Liang L, Yang X, Li Q, et al. Vibration damping of carbon nanotube assembly materials. Advanced Engineering Materials. 2018;**20**(3):1700647

[84] Zhang X, Zhi D, Sun L, Zhao Y, Tiwari MK, Carmalt CJ, et al. Super-durable, non-fluorinated superhydrophobic free-standing items. Journal of Materials Chemistry A. 2018;**6**(2):357-362

[85] Chen C-Y, Yoshiba M, Nagoshi T, Chang T-FM, Yamane D, Machida K, et al. Pulse electroplating of ultra-fine grained Au films with high compressive strength. Electrochemistry Communications. 2016;**67**:51-54

[86] Song C, He P, Lin T, Wei H, Yang W. Electroplating assisted diffusion bonding of ZrC–SiC composite for full ceramic joints. Ceramics International. 2014;**40**(5):7613-7616

[87] Wang B, Guo Z. Superhydrophobic copper mesh films with rapid oil/water separation properties by electrochemical deposition inspired from butterfly wing. Applied Physics Letters. 2013;**103**(6):63704

[88] Pham VH, Dickerson JH. Superhydrophobic silanized melamine sponges as high efficiency oil absorbent materials. ACS Applied Materials & Interfaces. 2014;**6**(16):14181-14188

[89] Wang C-F, Chen W-Y, Cheng H-Z, Fu S-L. Pressure-proof superhydrophobic films from flexible carbon nanotube/polymer coatings. Journal of Physical Chemistry C. 2010; **114**(37):15607-15611

[90] Wang J, Li Y, Kong Y, Zhou J, Wu J, Wu X, et al. Non-fluorinated superhydrophobic and micro/nano hierarchical Al doped ZnO film: The effect of Al doping on morphological and hydrophobic properties. RSC Advances. 2015;**5**(99):81024-81029

[91] Subramanian R, Shanmugam K, Marappan S. Fabrication of robust superhydrophobic coatings using PTFE-MWCNT nanocomposite: Supercritical fluid processing. Surface and Interface Analysis. 2018;**50**(4):464-470

[92] Sun Q, Aguila B, Verma G, Liu X, Dai Z, Deng F, et al. Superhydrophobicity: Constructing homogeneous catalysts into superhydrophobic porous frameworks to protect them from hydrolytic degradation. Chem. 2016;**1**(4):628-639

[93] Cai Y, Chang W, Luo X, Sousa AML, Lau KHA, Qin Y. Superhydrophobic structures on 316L stainless steel surfaces machined by nanosecond pulsed laser. Precision Engineering. 2018;**52**:266-275

[94] Schwiedrzik J, Raghavan R, Bürki A, LeNader V, Wolfram U, Michler J, et al. *In situ* micropillar compression reveals superior strength and ductility but an absence of damage in lamellar bone. Nature Materials. 2014;**13**(7):740-747

[95] Ali MB, Bessueille F, Chovelon JM, Abdelghani A, Jaffrezic-Renault N, Maaref MA, et al. Use of ultra-thin organic silane films for the improvement of gold adhesion to the silicon dioxide wafers for (bio)sensor applications. Materials Science and Engineering: C. 2008;**28**(5):628-632

[96] Hamel S, Tsukamoto T, Tanaka S, Fréchette LG. Microfabrication of a polymer based bi-conductive membrane for a polymer electrolyte membrane fuel cell. Journal of Physics Conference Series. 2013;**476**(1):12109

[97] Peng F, Wang D, Ma X, Zhu H, Qiao Y, Liu X. "Petal effect"-inspired superhydrophobic and highly adhesive coating on magnesium with enhanced corrosion resistance and biocompatibility. Science China Materials. 2018;**61**(4):629-642

[98] Crick CR, Parkin IP. A single step route to superhydrophobic surfaces through aerosol assisted deposition of rough polymer surfaces: Duplicating the lotus effect. Journal of Materials Chemistry. 2009;**19**(8):1074-1076

[99] Barahman M, Lyons AM. Ratchetlike slip angle anisotropy on printed superhydrophobic surfaces. Langmuir. 2011;**27**(16):9902-9909

[100] Xu L, Geng Z, He J, Zhou G. Mechanically robust, thermally stable, broadband antireflective, and superhydrophobic thin films on glass substrates. ACS Applied Materials & Interfaces. 2014;**6**(12):9029-9035

[101] Jiang W, He J, Xiao F, Yuan S, Lu H, Liang B. Preparation and antiscaling application of superhydrophobic anodized CuO nanowire surfaces. Industrial and Engineering Chemistry Research. 2015;**54**(27):6874-6883

[102] Lee C, Choi C-H, Kim C-J. Superhydrophobic drag reduction in laminar flows: A critical review. Experiments in Fluids. 2016;**57**(12):176

[103] Wang X, Ding B, Yu J, Wang M. Engineering biomimetic superhydrophobic surfaces of electrospun nanomaterials. Nano Today. 2011;**6**(5):510-530

[104] Rykaczewski K, Chinn J, Walker ML, Scott JHJ, Chinn A, Jones W. Dynamics of nanoparticle self-assembly into superhydrophobic liquid marbles during water condensation. ACS Nano. 2011;**5**(12):9746-9754

[105] Wang Z, Zhu H, Cao N, Du R, Liu Y, Zhao G. Superhydrophobic surfaces with excellent abrasion resistance based on benzoxazine/mesoporous SiO_2. Materials Letters. 2017; **186**:274-278

[106] Golovin K, Boban M, Mabry JM, Tuteja A. Designing self-healing superhydrophobic surfaces with exceptional mechanical durability. ACS Applied Materials & Interfaces. 2017;**9**(12):11212-11223

[107] Wang H, Xue Y, Ding J, Feng L, Wang X, Lin T. Durable, self-healing superhydrophobic and superoleophobic surfaces from fluorinated-decyl polyhedral oligomeric silsesquioxane and hydrolyzed fluorinated alkyl silane. Angewandte Chemie, International Edition. 2011;**50**(48):11433-11436

[108] Ramakrishna S, Kumar KSS, Mathew D, Nair CPR. A robust, melting class bulk superhydrophobic material with heat-healing and self-cleaning properties. Scientific Reports. 2015;**5**:18510

[109] Wu M, Ma B, Pan T, Chen S, Sun J. Silver-nanoparticle-colored cotton fabrics with tunable colors and durable antibacterial and self-healing superhydrophobic properties. Advanced Functional Materials. 2016;**26**(4):569-576

[110] Li Y, Ge B, Men X, Zhang Z, Xue Q. A facile and fast approach to mechanically stable and rapid self-healing waterproof fabrics. Composites Science and Technology. 2016;**125**:55-61

Chapter 3

Hydrophobic Surface Modification of Silk Fabric Using Plasma-Polymerized HMDSO

Bornali Sarma

Additional information is available at the end of the chapter

http://dx.doi.org/10.5772/intechopen.80304

Abstract

In this work, we study the wetting properties of silk fabrics by deposition of plasma-polymerized (PP) hexamethyldisiloxane (HMDSO) using low-pressure plasma-enhanced chemical vapor deposition (PECVD). Recently hydrophobic properties are under active research in textile industry. The effect of exposure time and power on the HMDSO-coated silk fabrics has been investigated. Water contact angle of PP-HMDSO-coated silk fabric surface has been measured as the function of power and coating time. Fabric surface has shown enhancement in hydrophobicity after coating. Attenuated total reflectance-Fourier transform infrared spectroscopy (ATR-FTIR) reveals the surface chemistry, and scanning electron microscopy (SEM) shows the surface morphology of the untreated and HMDSO-coated fabrics, respectively. In the case of untreated fabric, water droplet absorbs swiftly, whereas, in the case of HMDSO-coated fabric, water droplet remains on the fabric surface with a maximum contact angle of 140°. The HMDSO-deposited silk surface is found to be durable after detergent washing. Common stains like ink, tea, milk, turmeric, and orange juice are tested on the surface of both fabrics. In HMDSO-coated fabrics, all the stains are bedded like ball droplet. The fabric is tilted to 45° angle; stain droplets easily roll off from the fabric.

Keywords: plasma polymerization, HMDSO, silk fabrics, hydrophobic, PECVD

1. Introduction

Silk is a natural protein fiber from silk cocoon. It is highly praised as the queen of textiles because of its properties such as softness, glossy appearance, wearer comfort, warmth, and biodegradability. Silk fibers have large numbers of polar groups such as -OH, -COOH, and $-NH_2$ which are the backbone and side chains of polypeptide molecules. These hydrophilic structures provide a great atmosphere for growth of bacteria and fungi [1]. Silk fibers are susceptible to environmental circumstances, such as sunlight, staining from dirt, and debris. It is essential to extend silk fabrics with water-repellent functional properties having great potential in stain-free textile products. It prevents from accidental staining or water damage. The hydrophobicity of the fabric surface depends on its chemical functional groups, surface energy, and physical geometry [2]. There are some wet chemical methods available to change the surface properties of the fabrics, based on solvent-borne treatment with alkyl or partially fluorinated alkyl components [3, 4]. Several studies showed that properties of fabrics could be altered through surface modification. Alternating the surface properties of natural silk fibers by deposition of fluorinated polymers on the surface of the fabrics, it is becoming hydrophobic in nature [5]. Iriyama and Yasuda et al. reported that plasma treatment of CF_4 and C_2F_6 did not give good durability on the surface of polymer [6–8]. The tensile strength and hydrophobicity of Muga silk fiber have been reported by using RF argon (Ar) plasma treatment [9]. Li and Jinjin increased the contact angle of silk fabric up to 120° by C_3F_6 plasma treatment [10, 11]. Silk fabrics treated with SF_6 plasma showed that F replaced H and fluorination improved the hydrophobic property of the samples [7, 12–14]. Teli et al. improved the hydrophobic property of silk and cotton fabric using atmospheric pressure plasma in the presence of helium-fluorocarbon gasses, He/1,3-butadiene, and He/dodecyl acrylate [15–17]. The hydrophobicity has been achieved by plasma sputtering of polytetrafluoroethylene (PTFE) [18]. Fluorine-based polymers have hydrocarbons that break up into toxic compounds of perfluorooctane sulfonic acid (PFOS) and perfluorooctanoic acid (PFOA). The US Environmental Protection Agency (EPA) banned the long-chain fluorocarbon materials (PFOS/PFOA) due to their toxicity to human and environment [19–23]. However, the use of fluorocarbon and partially fluorinated alkyl compounds is undesirable due to the potential risks of the degradation by-products to human health and the environment, exceptionally high greenhouse effect compared to CO_2 [24, 25]. Many researchers' industrialized nanocoatings, like ZnO_2, Cu, TiO_2, DLC, etc., on the fabrics to increase the hydrophobic properties are still under active research [26–32].

Plasma-enhanced chemical vapor deposition (PECVD) of silicon compounds is an eco-friendly process and has been used to deposit ultrathin layer on the fabrics at a lower temperature without causing any thermal damages [33]. PECVD employs the conversion of monomer into reactive radicals, ions, and neutral molecules. Plasma polymerization creates a polymer film of organic compounds that do not polymerize under normal chemical polymerization process because it involves electron impact dissociation and ionization for chemical reactions [34]. Moreover, to reduce waste, pollution, water, energy, and time, plasma polymerization technology is employed, and it is a dry clean process which does not affect the environment. Silicon containing precursors like tetramethylsilane (TMS), tetraethoxysilane (TEOS), and hexamethyldisiloxane (HMDSO) is used for surface modification in textile industries. Among these, HMDSO precursor is nontoxic and nonexplosive and has a high vapor pressure at room temperature than TEOS [35–38]. Due to the presence of methyl groups; Si, H, and C atoms; and oxygen bond on the HMDSO, it changes the surface of the fabric into hydrophobic [35–39]. PP-HMDSO along with inert gasses in various natural as well as synthetic and blended textile substrates such as cotton, polyester wool, polypropylene, etc. has been studied by various researchers. Riccardi et al. observed that HMDSO-air was deposited by using dielectric barrier discharge plasma on silk surface by atmospheric pressure plasma to obtain a water-repellent

http://dx.doi.org/10.5772/intechopen.80304

silk fiber [40]. Hocker performed a deposition of HMDSO on oxygen-treated cotton fabric and achieved a maximum contact angle of 130° [41]. Ji et al. obtained the hydrophobic properties of polyester fiber by making use of an in-line atmospheric RF glow discharge plasma in a mixture of Ar and HMDSO [42, 43]. Palaskar et al. reported that HMDSO was coated with the mixture of helium and argon carrier gas for generating dielectric barrier discharge plasma on the polyester-/cotton-blended fabrics and improved the wetting properties of the fabrics [44]. Kale and Palaskar examined that deposition of TESO and HMDSO precursors by using PECVD was carried out on nylon 66 fabrics and HMDSO deposition rendered more hydrophobicity than TEOS [33]. Plasma polymer thin layers were deposited from pure HMDSO on polyimide substrate for water-repellent property enhancement and charge storage stability [45]. Shahidi et al. observed that HMDSO/N_2 plasma polymerization of wool fabrics improves anti-felting properties and dyeing behavior [39]. Plasma polymerization of organosilicon compounds is used in textile materials to increase functional properties of the materials. A lower surface energy indicates higher contact angle and greater hydrophobicity. A deposition of pure HMDSO by PECVD at low pressure on the silk fabric has not been reported yet.

In this study, the plasma polymerization of HMDSO on silk fabric has been performed by using the PECVD method at low pressure without causing thermal damage to the fabrics. PECVD coating technology has many advantages over conventional wet chemical methods. It has been used in a variety of deposition applications at a lower temperature. PP-HMDSO coating gives the possibility to obtain durable water-repellent surface due to retention of methyl groups. This coating makes the silk fabric surface water resistant, preventing it from accidental staining or water damage.

2. Materials and methods

2.1. Materials

A pure degummed silk fabric is purchased from a silk center with a warp count of 38 per/cm and a weft count of 38 per/cm 100% pure hexamethyldisiloxane (HMDSO), and the molecular formula is $C_6H_{18}OSi_2$. It is a colorless and highly volatile liquid, and the chemical structure is shown in **Figure 1**. It is obtained from Sigma-Aldrich.

2.2. Experimental

Plasma polymerization of HMDSO coating has been deposited onto silk fabric, and the experiment is carried out in a capacitively coupled plasma-enhanced chemical vapor deposition technique (PECVD) using pure HMDSO as a liquid precursor. PP-HMDSO depends on system pressure, discharge power, and coating time. The plasma reactor composed of stainless steel process chamber (24 cm height and 60 cm diameter) powered by a radio frequency generator (13.56 MHz) at room temperature. The chamber is evacuated to a base pressure of 1×10^{-5} mbar using rotary and diffusion pump. Plasma reactor consists of a pair of parallel symmetrical electrodes (35 cm diameter) separated by a distance of 3.5 cm. The schematic experimental setup is shown in **Figure 2**. The upper electrode has multipoint gas feeding

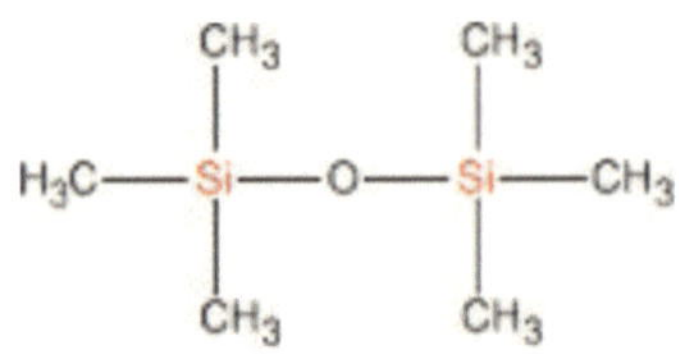

Figure 1. Chemical structure of hexamethyldisiloxane.

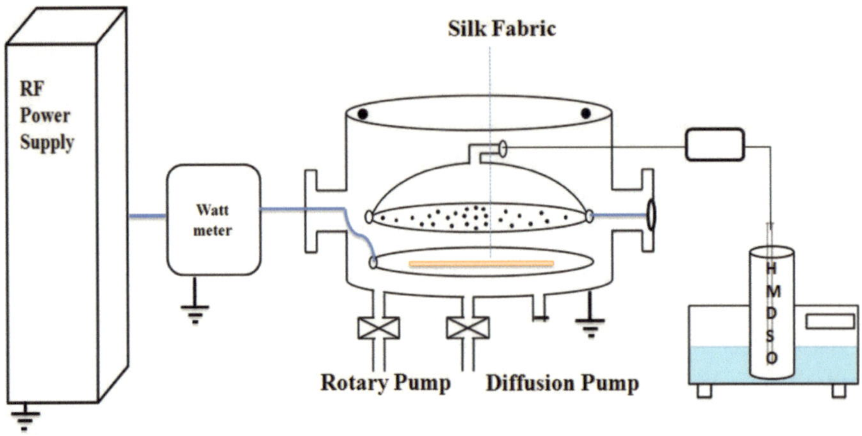

Figure 2. Experimental setup.

Sample code	Operating power (W)	Exposure time (min)	Pressure (mbar)
a	—	—	—
b	100	7	1.5×10^{-1}
c	100	15	1.5×10^{-1}
d	150	7	1.5×10^{-1}

Table 1. Operating parameters of PP-HMDSO.

shower head which is electrically grounded along with the chamber wall. The HMDSO is heated to 60°C using a constant temperature water bath. The working pressure is maintained at 0.15 mbar, and vaporized HMDSO liquid precursor is injected into the reactor using a needle valve without any carrier gas. The samples are placed on the surface of the lower electrode. The deposition is carried out for 7 min at 100 W and 150 W power and for 15 min at only 100 W RF power. The detailed experimental parameters of PP-HMDSO are reported in **Table 1**.

2.3. Instrumentations

FTIR spectra of the untreated and HMDSO-coated silk fabric are recorded by ATR-FTIR (Nicolet 6700, Thermo Scientific, USA). Morphologies of silk fabric, HMDSO-deposited fabrics, are examined with a scanning electron microscope (LEO s440i) at 10 kV and a magnification of 500×,

5000×, and 10,000×. Prior to SEM examination, the samples are pre-coated with gold sputtering to prevent charging of the samples by the electron beam. The static water contact angles on the untreated silk fabric and HMDSO-coated fabrics are measured using video contact angle optima (AST Products, Inc.) goniometer. A 5 μL drop of deionized water is placed on the substrate. The values of the static contact angles reported are the average of three measured values. Wet-out time is measured according to AATCC Test 79-1995. A 0.1 ml distilled water droplet is allowed to fall from a height of 5 cm onto the surface, and the time required for the droplet to be fully absorbed by the fabric is taken as the wet-out time. To study the durability of the HMDSO-coated fabric, washing test has been conducted. Based on the home laundering procedure, the HMDSO-coated fabrics are soaked in 60 ml of a 5.0% aqueous home laundering Surf Excel detergent (sodium carbonate, sodium aluminosilicate, alcohol ethoxylate, and sodium perborate monohydrate). The fabric is soaked in the detergent solution for 30 min at room temperature and rinsed with distilled water for several times and dried at room temperature overnight. Water repellency spray tester of AATCC Test method 22-2005 is modified to test the self-cleaning ability. In order to study the aging effect of HMDSO-coated silk fabrics, it has been left at room temperature for 100 days. Photographic images of different stains like ink, tea, milk, turmeric, and orange juice droplets on silk fabrics with and without HMDSO coating were measured by a digital camera after 100 days.

2.4. Results and discussion

2.4.1. ATR-FTIR

Attenuated total reflectance-Fourier transform infrared spectroscopy (ATR-FTIR) is used to examine functional groups of the untreated and HMDSO-coated silk fabrics and is shown in **Figure 3**. From **Figure 3a**, it is seen that the ATR-FTIR spectra of untreated silk fabric, the IR spectral region from 1700 to 1500 cm^{-1}, are due to peptide backbone, and the characteristic bands at 1621 cm^{-1} (amide I) are due to β-sheet confirmation of C=O stretching vibrations,

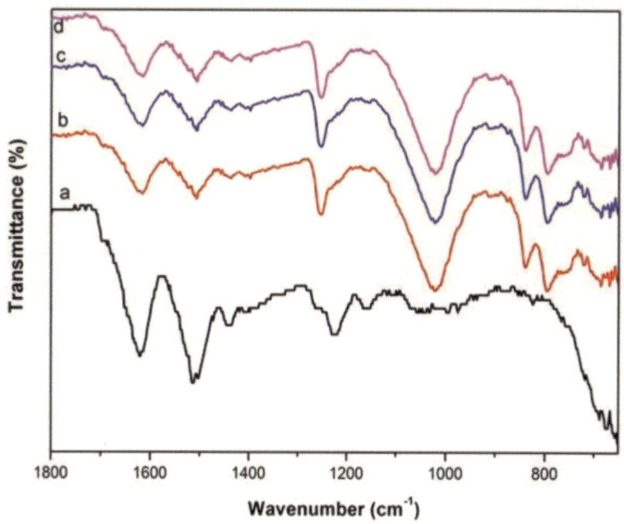

Figure 3. ATR-FTIR of (a) uncoated, (b) 100 W_7 min, (c) 100 W_15 min, (d) 150 W_7 min.

while band at 1514 cm^{-1} (amide II) is assigned due to the random coil conformation of fibroin molecules. The peak that appeared at 1226 cm^{-1} (amide III) is due to C—N stretching vibrations of the secondary structure of β-sheet [46]. **Figure 3b–d** shows the HMDSO-coated silk fabric; it is very clear that additional peak at 1218 cm^{-1} is due to symmetric bending of $Si\text{-}CH_3$. A band at 1041 cm^{-1} is exhibiting due to Si—O—Si stretching vibrations, whereas the peaks appeared at 970 and 763 cm^{-1} are due to Si-C rocking vibrations in the $Si—CH_3$ groups. The change in the peak intensities of the region from 1400 to 1800 cm^{-1} is attributed to the deposition of HMDSO [44].

2.4.2. SEM

SEM images of untreated and HMDSO-coated silk fibers are shown in **Figure 4**. **Figure 4a1–a3** shows SEM images of the untreated silk fabric at a magnification of 500×, 5000×, and 10,000×, respectively. It is clearly observed that the untreated silk fiber has a tranquil surface and is free from harshness. Since the surface of the silk is smooth, it has high polar groups which absorb the polar group immediately. **Figure 4b1–b3** and **c1–c3**shows the HMDSO deposition at operating power 100 W and exposure time of 7 and 15 min with a magnification of 500×, 5000×, and 10,000×, respectively. It is seen in the HMDSO-coated silk fabric SEM images (**Figure 4b1**) that the presence of some flakes like structure and roughness of the silk surface (**Figure 4b2**) is altered in the case of pop-HMDSO silk compared to untreated silk surface.

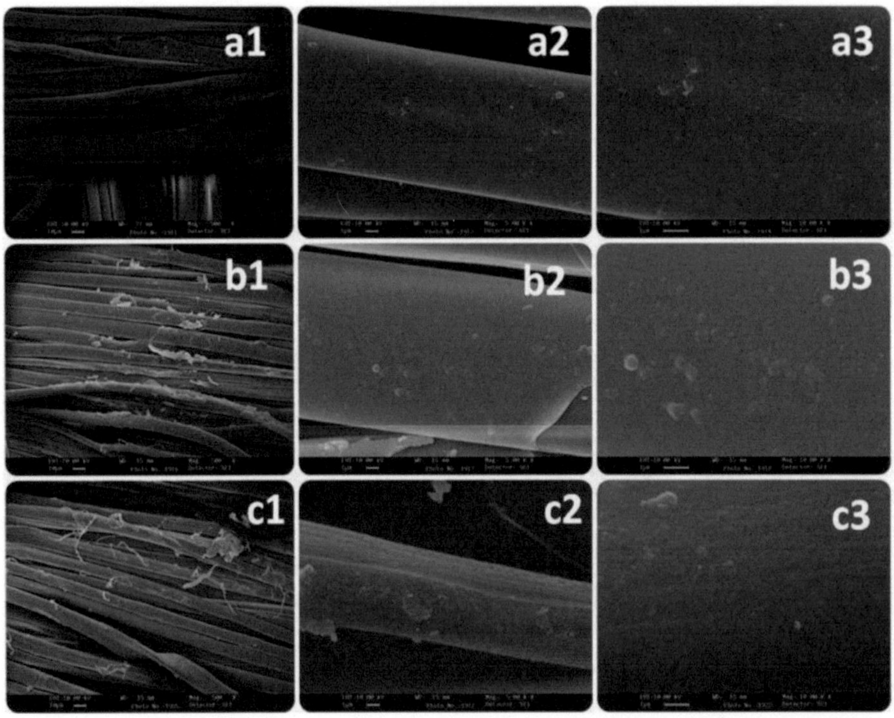

Figure 4. SEM images of (a) uncoated, (b) 100 W_7 min, (c) 100 W_15 min.

http://dx.doi.org/10.5772/intechopen.80304

Similarly **Figure 4c1** shows the presence of a thin layer on the surface of the fabrics which would render the attachment of HMDSO, and **Figure 4c2** shows the higher magnification of fabrics with increased surface roughness. However, by increasing the deposition time at constant power, the deposition of HMDSO increases the surface roughness and promotes the hydrophobicity properties.

2.4.3. Contact angle

Contact angle measurements are carried out using 5 μl water as the polar liquid. The untreated silk fabric has rich hydroxyl, carboxyl, and amine groups on its surface which exhibit a hydrophilic nature. The contact angles and wet-out time results are listed in **Table 2**.

When a thin layer of HMDSO is coated on the surface of the fabric, the fabric becomes hydrophobic due to the presence of $(Si{-}CH_3)_3$ rocking, Si—O—Si, and $Si{-}CH_3$. **Figure 5** shows

Operating power (W)	Exposure time (min)	Wet-out time (s)	Contact angle (°)
—	0 (Untreated)	44	—
100	7	>3600	134
100	15	>3600	135
150	7	>3600	140

Table 2. Contact angle and wet-out time of silk fabrics coated by HMDSO.

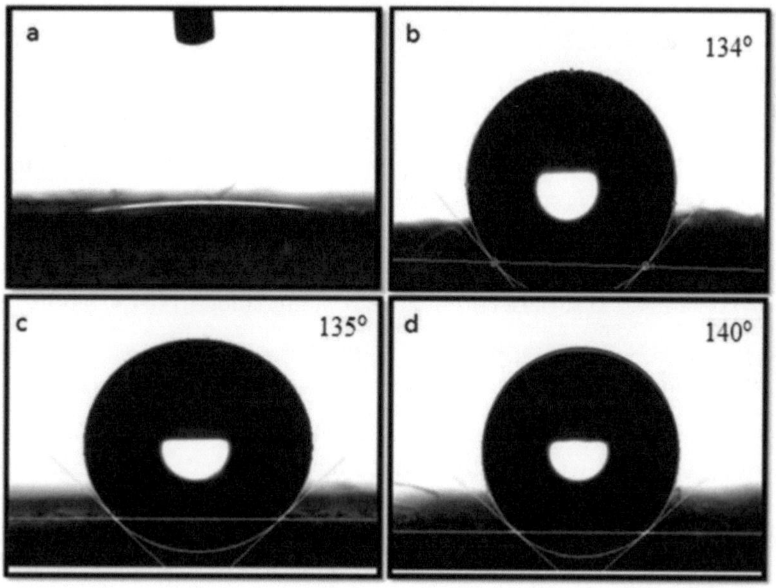

Figure 5. Static water contact angles of (a) uncoated, (b) 100 W_7 min, (c) 100 W_15 min, (d) 150 W_7 min.

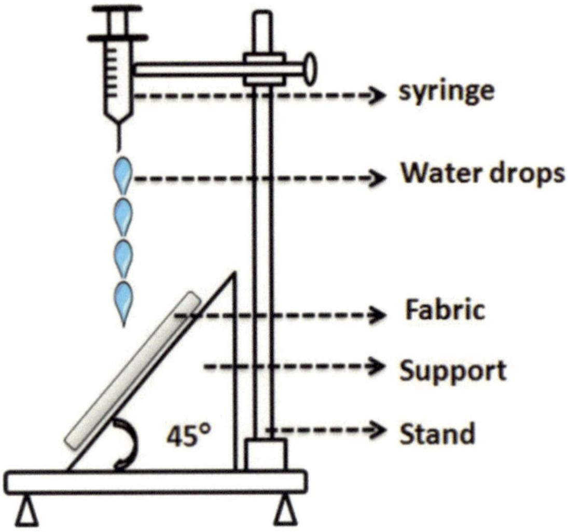

Figure 6. Experimental setup of the self-cleaning test.

a 5 μl water droplet on the silk fabric. It is seen that there is a drastic increase in the surface contact angle from 0° to 140° for untreated and HMDSO-coated silk fabric. **Table 2** represents the contact angle of HMDSO-coated samples for an exposure time of 7 min with the effect of discharge power 100 W (b) and 150 W (d), respectively. It is evident from the contact angle that more deposition is observed in samples treated with 150 W and its contact angle is 140°. From the above discussion, it shows that lower treatment time (7 min) and higher discharge power (150 W) result in more deposition on the fabric. However, there are no significant changes in the wet-out time of all HMDSO-coated fabrics.

2.4.4. Self-cleaning ability test

Water repellency spray tester of AATCC Test method 22-2005 is modified to test the self-cleaning ability [47]. **Figure 6** shows the experimental setup of the self-cleaning ability of the fabric. Graphite powder is spread on the HMDSO-coated fabrics. The fabric is tilted and placed at the center of the tester on a 45°-angle slope, and by using the syringe, water drop is allowed to fall on the fabric surface from a distance of 150 mm. The water droplets easily roll off along with contaminated surface and remove contamination from the fabric.

2.4.5. Home laundering and aging effect

To investigate the durability of HMDSO-coated silk fabric, the fabrics are washed with detergents. After detergent washing, there is a slight decrease in the contact angle, which will not affect the hydrophobic properties of the fabric. The aging effect of HMDSO-coated fabrics reveals that there are no significant visual changes in the wettability of silk fabrics after 100 days of deposition of the coating, which suggests good durability of the treatment. **Figure 7a** shows the common stains like ink, tea, milk, turmeric, and orange juice tested on the surface of both fabrics. In HMDSO-coated fabrics, all the stains are bedded like ball droplet.

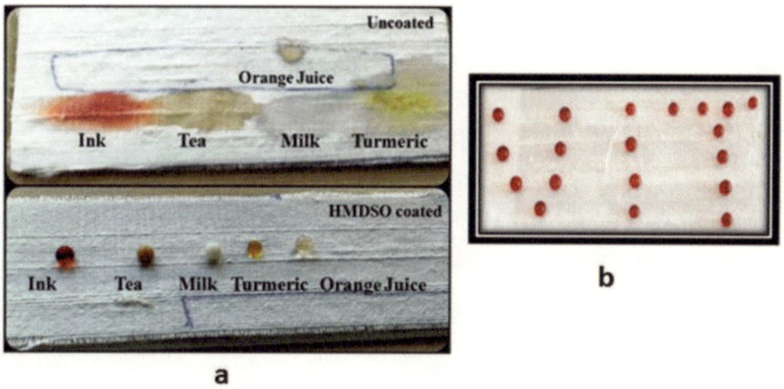

Figure 7. Photograph of stain with different kinds of liquids on the uncoated and HMDSO-coated fabric after 100 days.

In the case of untreated fabric, it's fully observed and stains the silk fabrics. **Figure 7b** shows an aging effect after 100 days, the ink droplet on the fabric bedded on the surface of the fabric.

3. Conclusion

The enhancement of hydrophobicity of the silk fabric has been achieved by deposition of pure HMDSO using plasma-enhanced chemical vapor deposition technique. The hydrophobic property has a great potential in the textile industry for stain-free fabrics. It is evident from the contact angle that more deposition is observed for a shorter exposure time along with higher discharge power (150 W) and its contact angle is 140°. Functional groups of HMDSO coating have been confirmed by ATR-FTIR spectroscopy; SEM images show the altered surface morphology of PP-HMDSO-coated silk fabric. Contact angle measurements reveal that the surface of silk fabric becomes more hydrophobic after deposition of PP-HMDSO. Coated fabrics are capable of repelling most aqueous liquids and dirt particles.

Acknowledgements

Authors are thankful to Plasma Science Society of India (PSSI) for providing financial support under PSSI fellowship and also thankful to the FCIPT for providing experimental facilities to carry out the research work.

Author details

Bornali Sarma

Address all correspondence to: bornali.sarma@vit.ac.in

Physics Division, School of Advanced Sciences, VIT University, Chennai, Tamil Nadu, India

References

[1] Chen F, Liu X, Yang H, Dong B, Zhou Y. A simple one-step approach to fabrication of highly hydrophobic silk fabrics. Applied Surface Science. 2016;**360**:207

[2] Ji YY, Kim SS, Kwon OP, Lee SH. Easy fabrication of large-size superhydrophobic surfaces by atmospheric pressure plasma polymerization with non-polar aromatic hydrocarbon in an in-line process. Applied Surface Science. 2009;**225**:4575

[3] Li Z, Xing Y, Dai J. Superhydrophobic surfaces prepared from water glass and non-fluorinated alkylsilane on cotton substrates. Applied Surface Science. 2008;**254**:2131

[4] Colleoni C, Donelli I, Freddi G, Guido E, Migani V, Rosace G. A novel sol-gel multi-layer approach for cotton fabric finishing by tetraethoxysilane precursor. Surface and Coating Technology. 2013;**235**:192

[5] Suanpoot P, Kueseng K, Ortmann S, Kaufmann R, Umongno C, Nimmanpipug P, et al. Surface analysis of hydrophobicity of Thai silk treated by SF6 plasma. Surface and Coating Technology. 2008;**202**:5543

[6] Iriyama Y, Yasuda T, Cho DL. Plasma surface treatment on nylon fabrics by fluorocarbon compounds. Journal of Applied Polymer Science. 1990;**39**:249

[7] Poletti G, Orsini F, Raffaele-Addamo A, Riccardi C, Selli E. Cold plasma treatment of PET fabrics: AFM surface morphology characterisation. Applied Surface Science. 2003;**219**:3

[8] Gorjanc M, Jazbec K, Šala M, Zaplotnik R, Vesel A, Mozetič M. Creating cellulose fibres with excellent UV protective properties using moist CF4 plasma and ZnO nanoparticles. Cellulose. 2014;**21**:3007

[9] Gogoia D, Choudhurya AJ, Chutiaa J, Pala AR, Dassa NN, Devib D, et al. Enhancement of hydrophobicity and tensile strength of muga silk fiber by radiofrequency Ar plasma discharge. Applied Surface Science. 2011;**258**:126

[10] Li S, Jinjin D. Improvement of hydrophobic properties of silk and cotton by hexafluoropropene plasma treatment. Applied Surface Science. 2007;**253**:5051

[11] Virk RK, Ramaswamy GN. Plasma and antimicrobial treatment of nonwoven fabrics for surgical gowns. Textile Research Journal. 2004;**74**:1073

[12] Chaivan P, Pasaja N, Boonyawan D, Suanpoot P, Vilaithong T. Low-temperature plasma treatment for hydrophobicity improvement of silk. Surface and Coating Technology. 2005;**193**:356

[13] Hodak SK, Supasai T, Paosawatyanyong B, Kamalangkla K, Pavarajarn V. Enhancement of the hydrophobicity of silk fabrics by SF6 plasma. Applied Surface Science. 2008;**254**:4744

[14] Selli E, Riccardi C, Massafra MR, Marcandalli B. Surface modifications of silk by cold SF6 plasma treatment. Macromolecular Chemistry and Physics. 2001;**202**:1672

[15] Teli MD, Samanta KK, Pandit P, Basak S, Gayatri TN. Hydrophobic silk fabric using atmospheric pressure plasma. International Journal of Bioresource Science. 2015;**2**:15

[16] Zhang J, France P, Radomyselskiy A, Datta S, Zhao J, Ooij WV. Hydrophobic cotton fabric coated by a thin nanoparticulate plasma film. Journal of Applied Polymer Science. 2003;**88**:1473

[17] Panda PK, Jassal M, Agrawal AK. Functionalization of cellulosic substrate using He/dodecyl acrylate plasma at atmospheric pressure. Surface and Coating Technology. 2013;**225**:97

[18] Huang F, Wei Q, Liu Y, Gao W, Huang Y. Surface functionalization of silk fabric by PTFE sputter coating. Journal of Materials Science. 2007;**42**:8025

[19] Moiz A, Vijayan A, Padhye R, Wang X. Chemical and water protective surface on cotton fabric by pad-knife-pad coating of WPU-PDMS-TMS. Cellulose. 2016;**23**:3377

[20] Palomares IR, Leganés F, Rosal R, Pinas FF. Toxicological interactions of perfluorooctane sulfonic acid (PFOS) and perfluorooctanoic acid (PFOA) with selected pollutants. Journal of Hazardous Materials. 2012;**201**:209

[21] Lau C, Butenhoff JL, Rogers JM. The developmental toxicity of perfluoroalkyl acids and their derivatives. Toxicology and Applied Pharmacology. 2004;**198**:231

[22] Zhang L, Niu J, Li Y, et al. Evaluating the sub-lethal toxicity of PFOS and PFOA using rotifer Brachionus calyciflorus. Environmental Pollution. 2013;**180**:34

[23] Wang W, Lockwood K, Boyd LM, Davidson MD, Movafaghi S, Vahabi H, et al. Superhydrophobic coatings with edible materials. ACS Applied Materials and Interfaces. 2016;**8**:18664

[24] Samanta KK, Johi AG, Jassal M, Agrawal AK. Study of hydrophobic finishing of cellulosic substrate using He/1, 3-butadiene plasma at atmospheric pressure. Surface and Coating Technology. 2012;**213**:65

[25] Zhao Q, Wu LYL, Huang H, Liu Y. Ambient-curable superhydrophobic fabric coating prepared by water-based non-fluorinated formulation. Materials and Design. 2016;**92**:541

[26] Xu B, Cai Z. Fabrication of a superhydrophobic ZnO nanorod array film on cotton fabrics via a wet chemical route and hydrophobic modification. Applied Surface Science. 2008;**254**:5899

[27] Dhineshbabu NR, Manivasakan P, Karthik A, Rajendran V. Hydrophobicity, flame retardancy and antibacterial properties of cotton fabrics functionalised with MgO/methyl silicate nanocomposites. RSC Advances. 2014;**4**:32161

[28] Shirtcliffe NJ, McHale G, Newton MI, Perry CC. Wetting and wetting transitions on copper-based super-hydrophobic surfaces. Langmuir. 2005;**21**:937

[29] Naebe M, Cookson PG, Rippon J, Brady RP, Wang X. Effects of plasma treatment of wool on the uptake of sulfonated dyes with different hydrophobic properties. Textile Research Journal. 2010;**80**(4):312

[30] Russaa MFL, Rovellaa N, Buergob MA, Belfiorec CM, Pezzinoc A, Criscia GM, et al. Nano-TiO_2 coatings for cultural heritage protection: The role of the binder on hydrophobic and self-cleaning efficacy. Progress in Organic Coatings. 2016;**91**:1

[31] Caschera D, Cortese B, Mezzi A, Brucale M, Ingo GM, Gigli G, et al. Ultra hydrophobic/superhydrophilic modified cotton textiles through functionalized diamond-like carbon coatings for self-cleaning applications. Langmuir. 2013;**29**:2775

[32] Intarasang W, Thamjaree W, Boonyawan D, Nhuapeng W. Effect of coating time on LPP treated silk fabric coated with ZnO_2 nanoparticles. Chiang Mai Journal of Science. 2013;**40**:1000

[33] Kale KH, Palaskar SS. Structural studies of plasma polymers obtained in pulsed dielectric barrier discharge of TEOS and HMDSO on nylon 66 fabrics. Journal of Textile Institute. 2012;**103**:1088

[34] Gaur S, Vergason G. Plasma polymerizatio: theory and practice. 43rd Annual Society of Vacuum Coaters Technical Conference Proceedings-Denver; 2000

[35] Hegemann D. Plasma polymerization and its applications in textiles. Indian Journal of Fibre and Textile Research. 2006;**31**:99

[36] Rombaldoni F, Mossotti R, Montarsolo A, Songia MB, Innocenti R, Mazzuchetti G. Thin film deposition by PECVD using HMDSO-O_2-Ar gas mixture on knitted wool fabrics in order to improve pilling resistance. Fibers and Polymers. 2008;**9**:566

[37] Zajikova L, Bursikova V, Perina V, Mackova A, Subedi D, Janca J, et al. Plasma modification of polycarbonates. Surface and Coatings Technology. 2001;**142**:449

[38] Vautrin-Ul C, Roux F, Boisse-Laporte C, Pastol JL, Chausse A. Hexamethyldisiloxane (HMDSO)-plasma-polymerised coatings as primer for iron corrosion protection: influence of RF bias. Journal of Materials Chemistry. 2002;**12**:231

[39] Shahidi S, Ghoranneviss M, Wiener J, Mortazavi H. Effect of Hexamethyldisiloxane (HMDSO)/ Nitrogen Plasma Polymerisation on the Anti Felting and Dyeability of Wool Fabric. Fibres and Textiles in Eastern Europe. 2014;**3**:116

[40] Höcker H. Plasma treatment of textile fibers. Pure and Applied Chemistry. 2002;**74**:423

[41] Riccardi C, Barni R, Esena P. Plasma treatment of silk. Solid State Phenomena. 2005;**107**:125

[42] Ji Y, Hong YC, Lee SH, Kim SD, Kim SS. Formation of super-hydrophobic and water-repellency surface with hexamethyldisiloxane (HMDSO) coating on polyethyleneteraphtalate fiber by atmosperic pressure plasma polymerization. Surface and Coatings Technology. 2008;**202**:5663

[43] Ji YY, Chang HK, Hong YC, Lee SH. Water-repellent improvement of polyester fiber via radio frequency plasma treatment with argon/hexamethyldisiloxane (HMDSO) at atmospheric pressure. Current Applied Physics. 2009;**9**:253

[44] Palaskar S, Kale kH, Nadiger GS, Desai AN. Dielectric barrier discharge plasma induced surface modification of polyester/cotton blended fabrics to impart water repellency using HMDSO. Journal of Applied Polymer Science. 2011;**122**:1092

[45] Ziari Z, Nouicer I, Sahli S, Rebiai S, Bellel A, Segui Y, et al. Chemical and electrical properties of HMDSO plasma coated polyimide. Vacuum. 2013;**93**:31

[46] Vyas SK, Shukla SR. Comparative study of degumming of silk varieties by different techniques. Journal of the Textile Institute. 2015:1

[47] Yun C, Islam MI, Lehew M, Kim J. Assessment of environmental and economic impacts made by the reduced laundering of self-cleaning fabrics. Fibers and Polymers. 2016;**17**:1296

Chapter 4

The Influence of Proteins Surface on the Ordering of Surrounded Water

Mateusz Banach, Leszek Konieczny and Irena Roterman

Additional information is available at the end of the chapter

http://dx.doi.org/10.5772/intechopen.80305

Abstract

Protein folding remains not satisfactory understood process. Considering the critical importance of water for proteins and other biologically active molecules, analysis of water-protein interactions should play a central role in studies concerning the folding process and biological activity of proteins. Folding simulations should acknowledge the aqueous solvent as an active partner which determines the final conformation of a protein. In the fuzzy oil drop model (which is applied in the presented analysis), the solvent is treated as a continuum—an external force field guiding the folding process. This interaction goes both ways: (1) the solvent shapes the protein and (2) the presence of a natively folded protein also alters the structure of the solvent (the structure of water has not heretofore been sufficiently studied—except for the solid state). This work focuses on this second reverse relationship, that is, the influence of proteins upon the structuralization of water. We formulate a hypothesis which is based on the fuzzy oil drop model. The ordering of the hydrophobic core which resides inside the protein and may include local discordances is analyzed from the point of view of its external effects. In accordance to the fuzzy oil drop model, the solvent is expected to "react" to local differentiation in the properties of the molecular surface. Our hypothesis remains speculative, since experimental studies have not yet yielded sufficient evidence to either prove or disprove it. The presented analysis bases on the assumption that a protein is nothing more than a tool engineered to perform a specific task. Thus, the protein's structure must encode its intended use and the inter-molecular communication system. Our study focuses on antifreeze proteins, which are particularly interesting since their function involves altering the properties of the solvent—specifically, preventing the formation of ice crystals.

Keywords: hydrophobicity, protein structure, hydrophobic core, water-protein surface interface, role of exposed hydrophobicity on protein surface

1. Introduction

Protein folding continues to attract a great deal of scientific interest in hopes of discovering its underlying mechanisms [1–6]. The search for computational algorithms is capable of reliably predicting the conformational properties of specific residue chain dates back to at least 1994, which is when the CASP challenge was launched [7]. That each residue sequence encodes a specific 3D structure is evident from the fact that protein folding—which continually occurs in living organisms—produces the same results each time [8, 9]. To-date protein folding models fail to acknowledge the involvement of the aqueous solvent, which plays a crucial role not only in protein folding, but also in other processes occurring in living cells. In molecular dynamics simulations, water is typically modeled as a set of molecules (expressing the known solvent density) [8, 10]. These molecules interact with polypeptide chain atoms in a pairwise (atom-atom) fashion.

In contrast, the fuzzy oil drop model which underpins the presented research treats the solvent as a continuum whose structural properties are unknown (for example, it is unclear why the density of water peaks at 4°C) but whose effects can be observed experimentally. The polar solvent causes hydrophobic residues to congregate at the center of the protein body, while hydrophilic residues are instead exposed on its surface. Nevertheless, hydrophobic residues are not perfectly isolated and can be detected on the surface of many proteins.

Rather than delve into the structural properties of proteins, the presented analysis focuses on the reverse relationship—the effect of the protein's presence upon the surrounding environment. This issue is important in light of the variable degree of ordering (or disorder) which characterizes the protein's hydrophobic core. More specifically, it refers to the occasional exposure of hydrophobicity on the surface and—by the same token—internalization of hydrophilic residues. Clearly, regardless of the structure of the solvent (which is treated as a continuum), exposure of hydrophobicity must result in local changes in its properties.

The assumption which forms the basis of the presented work is that the 3D structure of a protein represents a balance between the effects of internal force fields (pairwise interactions between atoms) and the external force field (the aqueous solvent, treated as a continuum).

In order to provide a mathematical model of the solvent, we refer to a 3D Gaussian form, which is assumed to represent an idealized (or "theoretical") distribution of hydrophobicity in a perfect protein molecule—with all hydrophobic residues internalized and all hydrophilic residues exposed on the surface. It turns out that while actual proteins do indeed conform to this model, they also exhibit certain deviations and localized discordances, which are associated with their function. For example, local exposure of hydrophobicity usually forms a suitable interface for protein-protein interactions [11–14], while local hydrophobicity deficiencies often characterize ligand/substrate binding cavities [15]. In addition to such localized effects (which can be formally quantified), the protein as a whole may diverge from the theoretical model by adopting an entirely different structural pattern, precluding the formation of a hydrophobic core. Such effects are observed, for example, in solenoid fragments, where instead of a centralized core we are faced with linear propagation of alternating bands of high and low hydrophobicity. This group also includes antifreeze proteins and pathological structures referred to as amyloids [16]. The difference between both groups is that antifreeze proteins contain—in addition to solenoid fragments—additional structural units which are locally accordant with the 3D Gaussian and provide the protein with solubility. Amyloids lack such structures and therefore remain insoluble [17, 18].

The presence of a protein in which the distribution of hydrophobicity is not in perfect agreement with environmental stimuli must exert an influence upon the environment itself. Exposure of hydrophobicity is "felt" by adjacent water molecules, which then react accordingly. This reverse relationship between a fully folded protein and the aqueous solvent is the focus of the presented study.

One of the major concepts in physical chemistry is the phenomenon of micellarization, producing various types of micelles (including spherical micelles) [16]. Bipolar molecules which comprise both hydrophobic and hydrophilic components aggregate to form structures which limit the entropically disadvantageous contact between hydrophobic fragments and water in favor of exposure of polar fragments [19–22].

The surfactant micelle, made up of identical loose units, exhibits high symmetry. This symmetry is additionally promoted by the large number of degrees of freedom characterizing each unit molecule—much like in the case of spherical or wormlike micelles [19–22].

The micelle may also intercalate external molecules, regardless of their size, if these molecules are capable of aligning themselves with the solvent without disrupting the overall symmetry of the system [23–25]. In all such cases, the surface of the micelle must be uniformly composed of polar groups, ensuring entropically advantageous interaction with water.

Of course, treating the protein as a "quasi" micelle, with properties similar to those exhibited by surfactant micelles, comes with certain caveats. The principal differences between proteins and surfactant micelles are twofold: first, in a residue chain, the distribution of hydrophobicity varies from amino acid to amino acid; second, the residues forming a polypeptide are not individual molecules—rather, they are connected with peptide bonds which significantly restrict their conformational freedom and therefore their ability to reach a location which would reflect their intrinsic hydrophobicity. Consequently, proteins do not follow the idealized distribution with perfect accuracy. Although it is, in principle, possible to design a sequence which would fold into a near-perfect spherical micelle, with excellent agreement between the theoretical and observed distribution of hydrophobicity [26], the vast diversity of biologically active proteins suggests that some proteins may be unable to generate a prominent central hydrophobic core. This, in turn, implies that the type and degree of local/global discordance versus the theoretical model are an expression of the quality referred to in biochemistry as "specificity". It should, however, be noted that in our study, the term does not refer to specificity of chemical interactions (e.g., between the protein and its ligand) but rather to the specific relation between the protein and the aqueous solvent, which is intimately linked to the existence and activity of numerous proteins.

The structural properties of water remain poorly understood—as evidenced by the lack of a convincing explanation of why the density of water peaks at 4°C. For this reason, we postulate extension of further experimental analysis of the aqueous solvent as a critical factor in mediating communication between molecules forming the solute. Further insight in this regard would help explain how the presence of water affects the protein—but also how the presence of proteins affects the solvent. The water environment shall also be treated as medium for inter-molecular communication. The characteristics of protein surface seem to play a critical role in this issue. This is why we attempt to demonstrate that the relationship water-protein is mutualistic.

2. Materials and methods

2.1. Data

The presented analysis concerns antifreeze proteins listed in **Table 1** (along with brief descriptions).

The study set presented in **Table 1** was assembled in an intentional manner. The protein—1J5B (type I antifreeze protein)—is a simple helix with highly discordant, however, very specific hydrophobic core. 2ZIB (type II antifreeze protein) exhibits minor discordance versus the model. The set is complemented by a globally discordant protein—multidomain antifreeze protein (5B5H) and a pathological (amyloid-2MVX) protein in which the distribution of hydrophobicity is entirely linear and consists of alternating bands of high and low hydrophobicity.

The discordances exhibited in each protein are quantified by applying methods described in the following subsection.

2.2. Fuzzy oil drop model

The fuzzy oil drop model has been thoroughly described in numerous publications with detailed presentation [31]. The authors also presented its application to antifreeze proteins [16] as well as to pathological proteins (amyloids) [17, 18]. For the purposes of the presented research, we will limit ourselves to a brief recapitulation of the model's core concepts, enabling the reader to understand the results presented further below.

PDB-ID	Protein	Chain length	Reference
Helix-1J5B	Antifreeze I type	37 aa	[27]
Globular-2ZIB	Antifreeze II type	130 aa	[28]
Solenoid-5B5H	Antifreeze-multidomain	223 aa	[29]
Amyloid-2MVX	Amyloid	73 aa/chain	[30]

Table 1. Set of proteins subjected to analysis, along with their basic properties and selected references.

At its heart, the fuzzy oil drop model is a modification of Kauzmann's original oil drop model [32]. That model likens the folding polypeptide to a drop of oil which—being a hydrophobic substance—attempts to limit the area of its contact with the aqueous solvent. Kauzmann divides the 3D protein structure into two layers: the internal (hydrophobic) layer and the external (polar) layer. The fuzzy oil drop model replaces this binary division with a continuous distribution where hydrophobicity peaks at the center and then progressively decreases along with the distance of the center, reaching almost 0 on the protein surface. This distribution can be mathematically expressed as a 3D Gaussian. If the protein molecule is encapsulated in a virtual ellipsoid (whose dimensions are adjusted to match the actual size of the protein), the Gaussian function directly yields the theoretical hydrophobicity values for arbitrary points within this capsule.

In mathematical terms, the 3D Gaussian is defined as follows:

$$\tilde{H}t_j = \frac{1}{\tilde{H}t_{sum}} \exp\left(\frac{-(x_j-\bar{x})^2}{2\sigma_x^2}\right)\exp\left(\frac{-(y_j-\bar{y})^2}{2\sigma_y^2}\right)\exp\left(\frac{-(z_j-\bar{z})^2}{2\sigma_z^2}\right) \quad (1)$$

$\bar{x}, \bar{y}, \bar{z}$ reflect the placement of the center of the ellipsoid (all three are equal to 0 at the origin of the coordinate system). σ coefficients are calculated as 1/3 of the greatest distance between any effective atom belonging the molecule and the origin of the system, once the molecule has been oriented in such a way that its greatest spatial extension coincides with a specific axis (for each axis separately).

The $1/Ht_{sum}$ coefficient ensures the normalization of both distributions (empirical and theoretical) and therefore enables comparative analysis. While theoretical hydrophobicity is defined at any point within the encapsulating ellipsoid, in practice, we are only interested in positions that correspond to effective atoms (averaged-out positions of all atoms comprising a given residue). Consequently, the sum has N components, where N is the number of residues in the chain. Each component is the theoretical value of hydrophobicity at the position of the given.

In contrast to the above, the actual distribution of hydrophobicity may diverge from theoretical values. Observed hydrophobicity (O) results from interactions between adjacent amino acids, which, in turn, depend on their mutual separation and the intrinsic hydrophobicity of each residue (which can be determined experimentally or on theoretical grounds [33]). Our analysis is based on the intrinsic hydrophobicity scale proposed in [31]. Under these assumptions, the observed hydrophobicity is given by the following formula [34]:

$$\tilde{H}o_j = \frac{1}{\tilde{H}o_{sum}} \sum_{i=1}^{N} (H_i^r + H_j^r) \begin{cases} \left[1 - \frac{1}{2}\left(7\left(\frac{r_{ij}}{c}\right)^2 - 9\left(\frac{r_{ij}}{c}\right)^4 + 5\left(\frac{r_{ij}}{c}\right)^6 - \left(\frac{r_{ij}}{c}\right)^8\right)\right] & \text{for } r_{ij} \le c \\ 0 \text{ for } r_{ij} > c \end{cases} \quad (2)$$

In both equations (Eqs. (1) and (2)), j denotes the position of the effective atom of the j-th residue. Ho_j is an aggregate value describing the interactions with neighboring residues (indexed i) at a distance not greater than 9 Å (this distance, c, is treated as the cutoff value for hydrophobic interactions, following the original model [34]). Applying a cutoff value implies that hydrophobic interactions are considered local and depend on the position of each residue. This function is empirically determined and, according to [34], expresses the force of hydrophobic interactions. H_i^r and H_j^r represent intrinsic hydrophobicity (constant for each residue) according to a predetermined scale, which can be arbitrary (in our study, the relevant scale is derived from [31]). r_{ij} is the distance between the i-th and the j-th residue, while N is the total number of residues in the chain.

Normalization of both distributions (with all T_i and all O_i adding up to 1.0) facilitates quantitative comparisons, as illustrated in **Figure 1**. Differences between both distributions may vary. **Figure 1A** shows a protein where the observed distribution is closely aligned with the theoretical distribution, while **Figure 1B** illustrates the opposite case—a significantly discordant protein.

Subjective assessment of the degree of discordance (cf. **Figure 1**) may be supplemented by quantitative analysis based on Kullback-Leibler's divergence entropy formula [35]:

$$D_{KL}(p|p^0) = \sum_{i=1}^{N} p_i \log_2(p_i/p_i^0) \quad (3)$$

where D_{KL} represents the distance between two distributions: "observed" (p) and "target" (p_0).

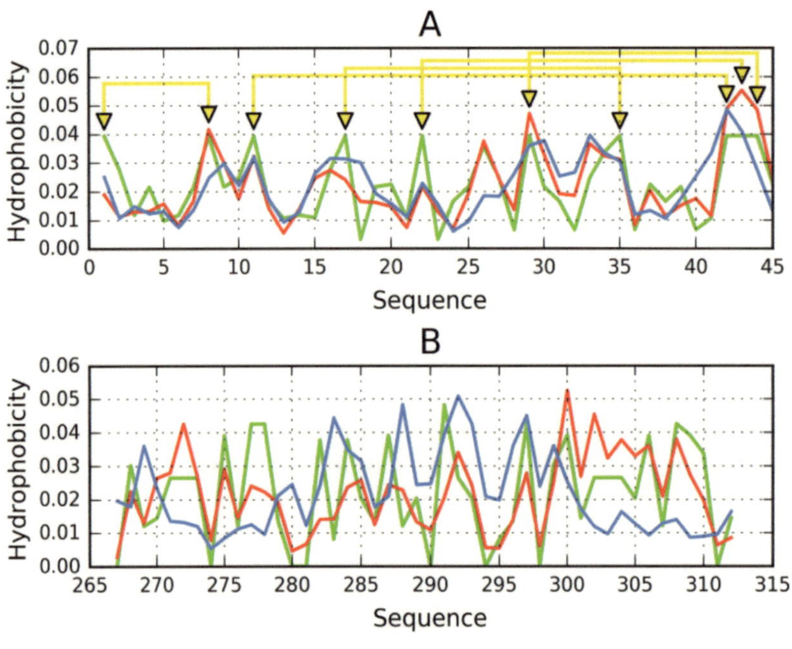

Figure 1. Examples of two proteins that differ with regard to their hydrophobic core structure: (A) accordant protein (5LAH-45 aa [35]) and (B) discordant protein (2MZ7-46 aa [36]). These proteins were selected to illustrate strong conformance/strong discordance, and are not part of the study set analyzed in this chapter. The disulphide bonds are shown as yellow lines.

To provide a quantitative measure of the differences between O and T, the latter distribution will be treated as a reference.

The status of a given protein is represented by D_{KL} values, which express its proximity to reference distributions treating distribution T as p^0 and distribution O as p:

$$O|T = \sum_{i=1}^{N} O_i \log_2(O_i/T_i) \tag{4}$$

The opposite reference distribution, R, represents the distribution deprived of any form of hydrophobicity differentiation (each residue represents equal hydrophobicity $R_i = 1/N$, where N is the number of residues in protein). The corresponding O | R relation is defined as follows:

$$O|R = \sum_{i=1}^{N} O_i \log_2(O_i/R_i) \tag{5}$$

For the examples illustrated in **Figure 1**, D_{KL} for protein (A) is 0.08, while for protein (B), its value is equal to 0.46.

These values cannot, however, be considered on their own, since D_{KL} is a relative measure of relative entropy (and moreover depends strongly on the number of residues in the chain). Introduction of two reference distributions allows the comparison: O | T larger than O | R suggests similarity between O and R distribution.

In order to avoid having to deal with two distinct values, we further introduce the so-called relative distance parameter, defined as follows:

$$RD = O|T/(O|T + O|R). \tag{6}$$

Values lower than 0.5 mean that O is a better match for T than for R. This is interpreted as the presence of a centralized hydrophobic core.

For the examples illustrated in **Figure 1**, RD values are 0.300 (A) and 0.680 (B), respectively.

It should be noted that various distributions may be used as reference. Besides R, we may also introduce another distribution, denoted H, which corresponds to the intrinsic hydrophobicity of each residue. In this case, the value of RD will express whether the given protein exhibits a distribution which more closely resembles T or H. Accordingly, we obtain two distinct values of RD: one for the T-O-R model and one for the T-O-H model. Such analysis becomes helpful when studying amyloid structures, which, according to FOD-based analysis, appear to be dominated by the conformational preferences of individual amino acids with no regard for the creation of a global hydrophobic core (unlike in globular proteins) [36]. To further underscore the influence of intrinsic hydrophobicity, we may also calculate correlation coefficients for three types of relationships: HvT, TvO, and HvO. We will soon show that strong discordance between O and T, where no centralized hydrophobic core can be observed, leads to negative HvT and TvO correlation coefficients, along with high positive values of the HvO coefficient.

2.3. The aqueous solvent—an interpretation rooted in the fuzzy oil drop model

Protein folding is regarded as the search for a global energy minimum. This implies optimization of the protein's internal force field. Nonbinding interactions (electrostatic, vdW, torsional potential, and others) are present in every molecule and produce structural forms which are optimal from the point of view of free energy minimization.

Classic protein folding algorithms (mostly based on molecular dynamics simulations) acknowledge the presence of the solvent by including a set number of external water molecules that interact with atoms belonging to the protein chain [12]. In contrast, the fuzzy oil drop model treats water as a continuum, represented by an external force field which directs hydrophobic residues toward the center of the molecule while exposing polar residues. This effect—next to the formation of disulfide bonds—is regarded as a primary force which stabilizes the protein's tertiary conformation.

The presented model therefore acknowledges the role of the solvent with no in-depth knowledge regarding the properties of this solvent: its mere presence is enough to drive the folding process, producing structures which are largely consistent with the 3D Gaussian distribution of hydrophobicity [26].

In fact, the structural ordering present in proteins is highly varied and may include local or global discordances. A local discordance manifests itself as either local excess hydrophobicity on the protein surface (providing a complexation interface for p-p interactions [11, 12]) or local deficient hydrophobicity inside the protein body (which often characterizes ligand or substrate binding pockets [13, 14]). On the other hand, global discordance occurs when the entire protein follows a different structural pattern which does not involve a centralized hydrophobic core—for example, linear propagation of alternating bands of high and low hydrophobicity, as observed in amyloid [17, 18].

Analysis of the observed distribution of hydrophobicity tells us whether the protein conforms to the theoretical model (and if it does—whether it includes any local deviations) or diverges from the model entirely.

The presented work focuses on the reverse phenomenon, that is, the influence of the protein upon its environment. Naturally, this is merely a postulate based on the observed nonalignment between O and T for many biologically active proteins. In such cases, the protein itself may be treated as the source of an external force field which acts upon the solvent. Its presence alone is sufficient to direct nearby water molecules. This phenomenon may potentially serve as a carrier of information between proximate proteins—a notion upon which the so-called iceberg hypothesis is based [37, 38].

One example which provides strong support for the above thesis involves antifreeze proteins. Such proteins are expected to work by disrupting the natural structuralization of water and thereby preventing formation of ice crystals. The explanation provided by our model contradicts older analyses, which search for ways in which nascent ice crystals might potentially dock to antifreeze proteins [39]. In our view, no such docking takes place. The docking model is also overly sensitive to concentrations of antifreeze proteins and does not explain the antifreeze effect observed in the context of small particles, for example, saccharides [39] or phospholipids [40]—not to mention individual ions, which are also observed to prevent formation of ice crystals in the macro scale.

3. Results

The effect of the presence of proteins upon the aqueous solvent is particularly evident in the case of antifreeze proteins [41]. Their task is to keep water from freezing in subzero temperatures, which would otherwise destroy cells and tissues, terminating all processes which the organism relies on in order to function. The increased mobility of water molecules on the surface of antifreeze proteins observed experimentally supports this expectations [42].

Organisms which undergo hibernation (including fish, plants, and other organisms) [41] have been found to produce specific proteins whose role is to counteract the formation of ordered ice crystals. Referring to calcium and sodium ions, if we assume that their presence disrupts the coordination of water particles (an effect exploited, e.g., for salting roads during winter), the same should be expected in the case of the aforementioned proteins.

The presented proteins have been intentionally selected in order to highlight various ways in which proteins affect the local structure of the aqueous solvent and counteract its tendency to crystallize.

3.1. Small (type I) antifreeze proteins

Type I antifreeze proteins are small and exhibit a uniform secondary structure, that is, they are entirely helical. One example is IJ5B, as listed in **Table 2**. **Figure 2** presents the theoretical and observed hydrophobicity distributions for this protein.

As evident in **Figure 2**, in place of the expected central hydrophobicity peak, we are faced with a near-sinusoidal pattern. Notably, the cyclical nature of this distribution does not correspond to individual twists which comprise the alpha helix (the number of residues per distribution cycle appears greater than 3). If this were the case, we would be dealing with a perfectly amphipathic helix, whereas the observed periodicity of changes results

Protein	RD		Correlation coeff.		
	T-O-R	T-O-H	HvT	TvO	HvO
1J5B	0.767	0.557	0.221	0.450	0.860
2ZIB	0.556	0.392	0.201	0.531	0.687
2ZIB No 70-78	0.498	0.361	0.214	0.572	0.693

Removal of the 70-78 fragment in 2ZIB produces a chain which is accordant with the theoretical distribution of hydrophobicity (position No 70-78).

Table 2. Small antifreeze proteins—fuzzy oil drop parameters.

in strong variations in the structural properties of water in close proximity of the protein. Hydrophilic patches likely attract water molecules resulting in an energetically optimal arrangement; however, in hydrophobic areas, the structuralization of water, while not precisely known, is most likely significantly altered (whatever the word "altered" means in this context). Water has been found to exhibit increased mobility close to antifreeze proteins, likely as a result of its complex interactions with the molecular surface [42]. It is postulated that the protein causes vortices to form in the surrounding medium (as illustrated in **Figure 3**).

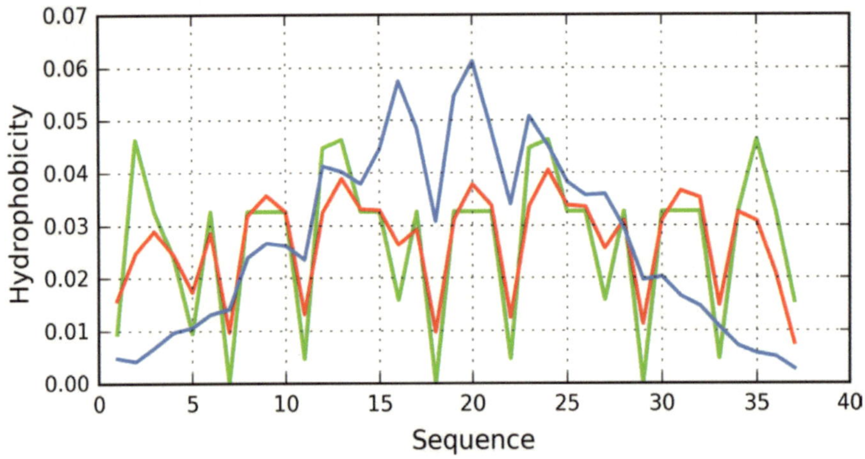

Figure 2. Type I antifreeze protein (IJ5B)—the figure illustrates the misalignemnts between T (blue) and O (red). The H distribution—green line.

Figure 3. The distribution of hydrophobic residues along the helix does not correspond to its structural periodicity. The image schematically depicts the expected reaction of the surrounding water particles, which are either attracted (teal) or repelled (red) by the molecular surface.

3.2. Small protein exhibiting marginal local discordance—2ZIB

2ZIB, with a chain length of 130 aa, provides an example of a type II antifreeze protein. Taken as a whole, it is characterized by RD slightly in excess of 0.5 (**Table 2** and **Figure 4**). Similar RD values in the T-O-R and T-O-H models, as well as comparable correlation coefficients, suggest that all factors (T, H, and O) represent some sort of consensus.

Figure 5 reveals the location of hydrophobic residues close to the surface. While most of the surface is composed of hydrophilic (polar) residues, local exposure of hydrophobicity produces a change in the structural properties of water as individual dipoles align themselves with the protein. The resulting structure disfavors ice crystal formation.

Removing the fragment at 70-78 (**Table 2**) produces a chain which largely conforms with theoretical expectations (95% of the remaining structure is accordant with the theoretical model). As noted, the presence of a charged surface induces changes in the structuralization of the aqueous solvent and thereby counteracts ice crystallization. It appears that the protein performs its function in much the same way as Na^{+}/Cl^{-} ions—with the added benefit of being able to expose a much larger surface area and thereby exert a more significant effect upon the surrounding medium.

The protein under consideration is also characterized by unexpectedly large number of disulfide bonds—in the chain of 130 aa, there are 5 SS-bonds. The disulfide bonds as well as the presence of hydrophobic core are treated as factors responsible for III-order stabilization in proteins. In this protein, the high stability is reached (and ensured) by both factors since the structure of hydrophobic core is almost perfect. Analysis of the role of disulfide bonds in relation to the structure of hydrophobic core was discussed in [43] revealing quite differentiated spectrum from high accordance with the hydrophobic core structure to clear opposite relation where the SS-bond fragments represent highly discordant (in respect to expected

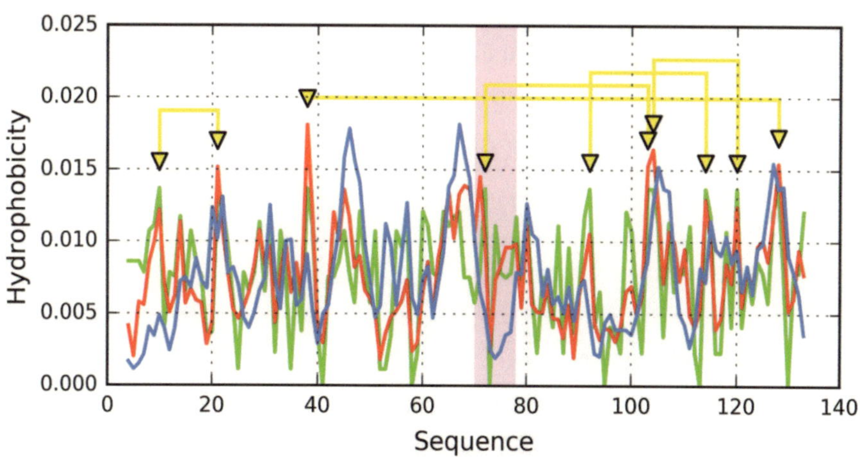

Figure 4. T and O distributions in a type II antifreeze protein (2ZIB). The main discordant section (exposure of hydrophobicity) has been marked on the horizontal axis. Disulfide bonds shown as yellow lines.

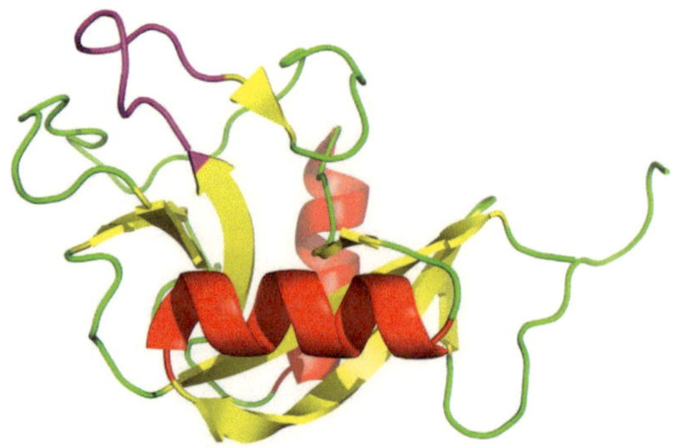

Figure 5. 3D (2ZIB) view with the 70-78 fragment highlighted in pink. This fragment represents excess hydrophobicity on the protein surface. Rest of the molecule represents the status accordant with 3D Gauss distribution of hydrophobicity.

one) formation of hydrophobic core. The presence of so many SS-bonds may additionally support stabilization of the structure as it is expected for antifreeze protein to be resistant to the influence of the surrounding.

3.3. Antifreeze protein which contains a solenoid fragment

5B5H provides an example of an antifreeze protein which contains a solenoid fragment. A general description of this protein can be found in [16]. It is classified as a multidomain protein (even though this is not reflected by the CATH [44, 45] specification stored in PDBsum [46, 47]). In this case, the concept of "multiple domains" appears to relate to the presence of diverse secondary structural motifs. T, O, and H distributions for this protein are visualized in **Figure 6**, revealing the fragments of variable alignment between profiles.

Table 3 presents the characteristics of this protein taking secondary fragments as individual units. The structure, taken as a whole, is regarded as discordant (**Figure 7A**). However some distinguished fragments appear to represent the distribution accordant with idealized distribution.

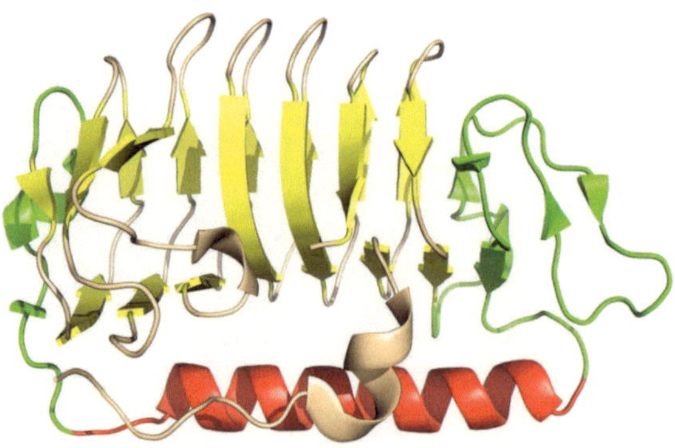

Figure 6. 3D structure of a multidomain antifreeze protein (5B5H). Fragments marked in color are discussed in details. The colors of fragments follow the colors used in **Figure 7**.

http://dx.doi.org/10.5772/intechopen.80305

5B5H	RD		Correlation coeff.		
FRAGMENT	T-O-R	T-O-H	HvT	TvO	HvO
Entire protein	0.676	0.538	0.386	0.436	0.754
N-stop 37-73	0.649	0.645	0.318	0.502	0.743
C-stop 98-113	0.304	0.093	0.298	0.810	0.421
Solenoid	0.669	0.526	0.386	0.436	0.754
Helix 74-98	0.389	0.431	0.551	0.835	0.688

Table 3. Quantitative description of 5B5H and its selected secondary structural components.

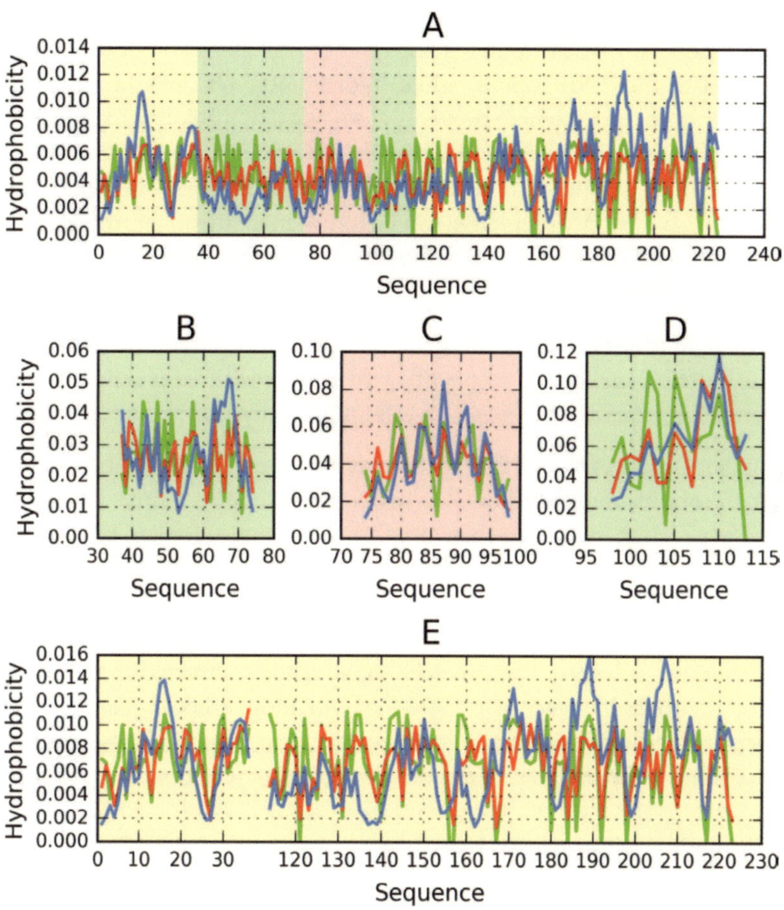

Figure 7. T (blue), O (red) and H (green) distributions for 5B5H. The profiles for colour highlighted fragments are shown individually in B, C and D. A—complete protein, B—N-terminal fragment, C—helix parallel to the solenoid, D—"stop" helix, E—solenoid fragment.

The solenoid (as observed in other examples of proteins) [16] represents the linear propagation of local maxima and minima of hydrophobicity. That is why to prevent the unlimited propagation, the "stop" fragments are necessary. Their status is expected to be accordant with the model. It means that their role is to allow water to penetrate and in consequence prevent uncontrolled grow of fibril.

N-terminal "stop" fragment seems not to play such role as its status is discordant (**Figure 7B**).

Helix—the status of the helix running along the solenoid is accordant with the fuzzy oil drop model, that is, its polar side is exposed to the solvent, while the hydrophobic side remains in contact with the solenoid (**Figures 6** and **7C**). It suggests its role to ensure the solubility of the entire molecule.

"Stop" helix C-terminal, the solenoid, by itself, is susceptible to unrestricted linear propagation, which theoretically may propagate in unlimited form (for example, by complexation of many solenoids). This undesirable effect is prevented by the presence of a short C-terminal helix (**Figures 6** and **7D**) which remains accordant with the theoretical model: polar residues are exposed to the environment, while hydrophobic residues face the solenoid. The helix therefore acts as a "cap", which prevents elongation of the solenoid. Notably, this is the most accordant fragment within the entire protein (as evidenced by its TvO correlation coefficient)—meaning that its conformation is driven by the tendency to generate a hydrophobic core. Similar "caps" (or "stop" fragments) can be found in many other proteins which include solenoid fragments, and the phenomenon may be exploited in designing drugs which arrest the propagation of amyloid fibrils [48].

Much like an amyloid fibril, the solenoid fragment is characterized by negative correlation coefficients for certain β-structural fragments, which suggests that it actively counteracts the natural tendency for the protein to generate a centralized hydrophobic core.

* Solenoid fragment (**Figures 6** and **7E**; marked in yellow) is the structural core of the protein. In this fragment, the observed distribution of hydrophobicity adopts a sinusoidal pattern with no distinct central peak. High accordance between O and H distribution can be seen. Additionally, the expected hydrophobicity concentration (high blue picks **Figure 7E**) is not present. Recurring fluctuations reflect both the symmetry of the solenoid itself and the arrangement of residues in the chain. Such alternating bands of high and low hydrophobicity are thought to affect the structural properties of the solvent—water molecules, which are attracted to hydrophilic patches but repelled by hydrophobic patches are also likely to adopt a linear "bandlike" structure. In [49], the authors suggest, and the experimentally prove, that water levitates above hydrophobic surfaces. This phenomenon, in turn, increases the mobility of water particles, as proved in [42].

3.4. Amyloid protein—Aβ structure (1-40)

Amyloids have a significant impact on the surrounding solvent. Here, we present the beta amyloid (1-40). This molecule meets the criteria specified in [17, 18] by exhibiting linear propagation of alternating bands of high and low hydrophobicity, which clearly deviate from the monocentric Gaussian in favor of a different structural pattern. The β-amyloid (1-40) structure [30] will be analyzed as a superfibril, as a protofibril and as well as an individual chain (component of the protofibril).

http://dx.doi.org/10.5772/intechopen.80305

3.4.1. Aβ(1-40) superfibril

The structure listed in PDB under the ID 2MVX comprises two individual protofibrils arranged symmetrically (C2 symmetry). Each protofibril resembles a flattened letter C. These protofibrils are connected by their tips, with their "backs" facing outward. **Figure 8** presents the theoretical and observed distributions of hydrophobicity for the superfibril when analyzed as a whole. Clearly, in place of a monocentric peak of hydrophobicity, we are instead dealing with a sinusoidal pattern expressed along the fibril's axis. This type of distribution is typical for amyloids and results from structural repeatability of the input chain, as well as from the symmetry between both protofibrils. In the T-O-R model, the RD value of the superfibril is 0.590, while in the T-O-H model, it is equal to 0.592, with correlation coefficients of 0.438, 0.673, and 0.727 for HvT, TvO, and HvO, respectively. These values suggest that the superfibril adopts a conformation which does not involve a central hydrophobic core. It moreover indicates that the resulting structure represents a compromise between the tendency to generate a hydrophobic core and the intrinsic properties of individual amino acids comprising the sequence.

The distribution illustrated in **Figure 8** reveals the expected (T) concentration of hydrophobicity in the central part of the fibril. The sinusoidal shape of the intrinsic (H) and observed (O) distribution curves is due to the previously postulated linear propagation of alternating bands of high and low hydrophobicity—a characteristic feature of amyloids [17, 18].

Figure 9 also reveals a repetitive pattern of alternating peaks and troughs, propagating along the chains. It should be noted that this single chart (red) represents several individual chains (shaped like a stack of sheets—on top of and beneath the sheet on which the chart is printed), each of which is characterized by identical arrangement of local maxima and minima. The variability seen in **Figure 9A** is related to the properties of edge chains which—being adjacent to only one other chain rather than two—exhibit slightly lower hydrophobicity than chains which make up the fibril's interior (**Figure 9B**). It is these interior chains which should be regarded as particularly representative for the amyloid form, owing to their capability for unrestricted propagation (observed both *in vitro* and *in vivo*).

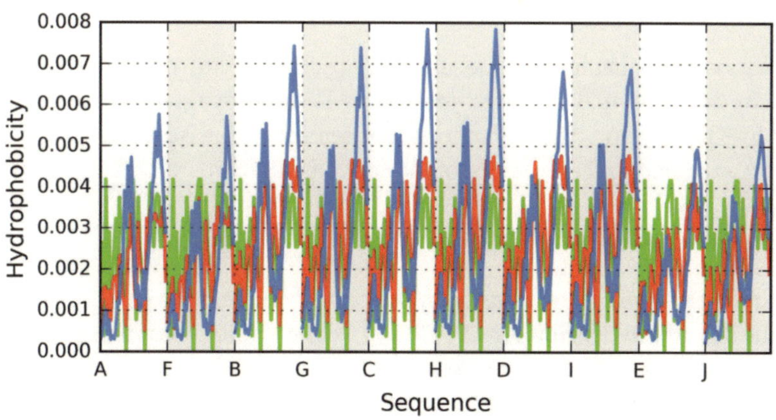

Figure 8. H (green), T (blue) and O (red) distributions for the Aβ(1-40) super-fibril (2MVX). White and gray backgrounds distinguish two proto-fibrils.

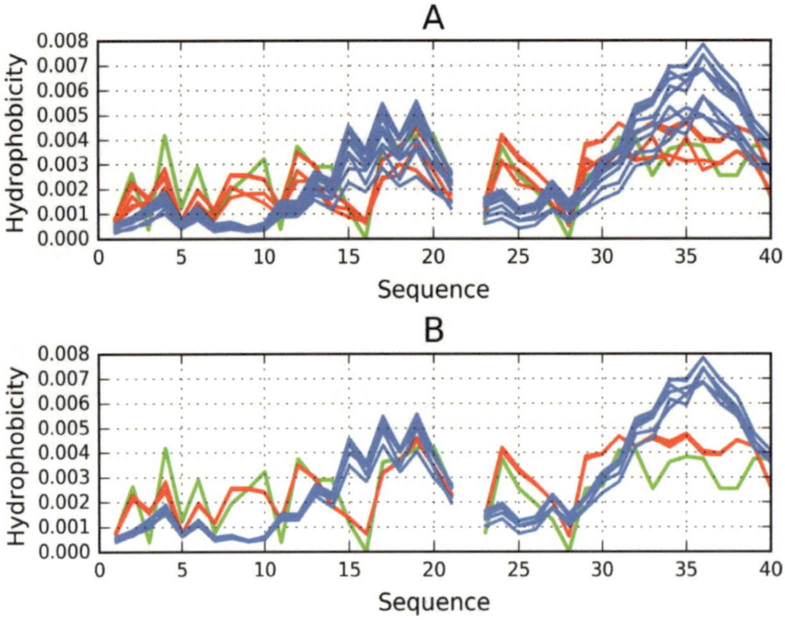

Figure 9. T (blue), H (green) and O (red) distributions for: A—all chains making up the super-fibril B—all chains except edge chains (AEFJ).

Plotting the theoretical 3D Gaussian for the entire complex (superfibril) enables us to assess the status of interface fragments, which include residues in contact with adjacent protofibrils. These residues have been analyzed in accordance with PDBSUM criteria [14]. The resulting RD values for the interface fragments are 0.432 and 0.387 for T-O-R and T-O-H, respectively, with correlation coefficients equal to 0.378, 0.672, and 0.658 for HvT, TvO, and HvO, respectively. These values suggest the status of the interface as accordant with the theoretical distribution of hydrophobicity as given by the fuzzy oil drop model, as long as the complex is analyzed as a whole (note the high values of TvO and HvO coefficients, indicating that the observed distribution is in agreement with both theoretical and the intrinsic distribution).

The above observations enable us to speculate that protofibrils are dominated by the intrinsic hydrophobicity of individual residues, leading to linear propagation of alternating bands of high and low hydrophobicity. In contrast, the complex (consisting of two protofibrils) is shaped by forces related to the presence of the aqueous solvent.

3.4.2. Structure of the Aβ(1-40) protofibril

Analysis of individual protofibrils has been performed on the basis of T, O, and H distributions, with the theoretical distribution (T) plotted for the single protofibril rather than for the entire complex. Each protofibril comprises five chains labeled A, B, C, D, and E. **Figure 10** illustrates the relevant distributions of hydrophobicity.

Again, we observe a repeating sinusoidal pattern instead of the expected central peak. The observed distribution is a result of the highly symmetrical arrangement of chains which form the protofibril and of their sequential identity (**Figure 10** and **Figure 11**).

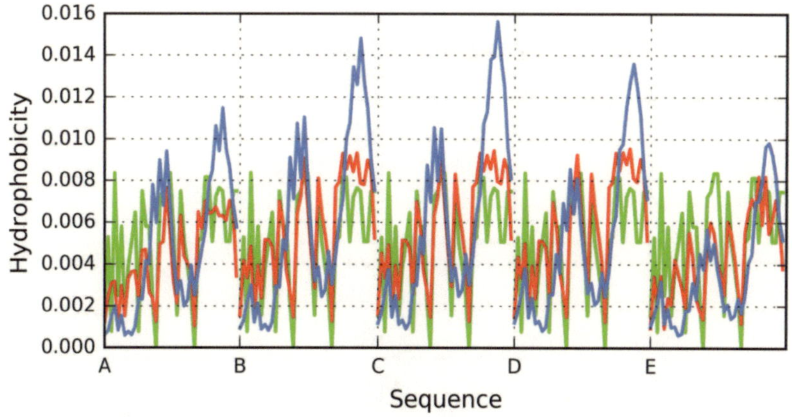

Figure 10. T (blue), O (red) and H (green) distributions for the Aβ(1-40) proto-fibril (chains ABCDE).

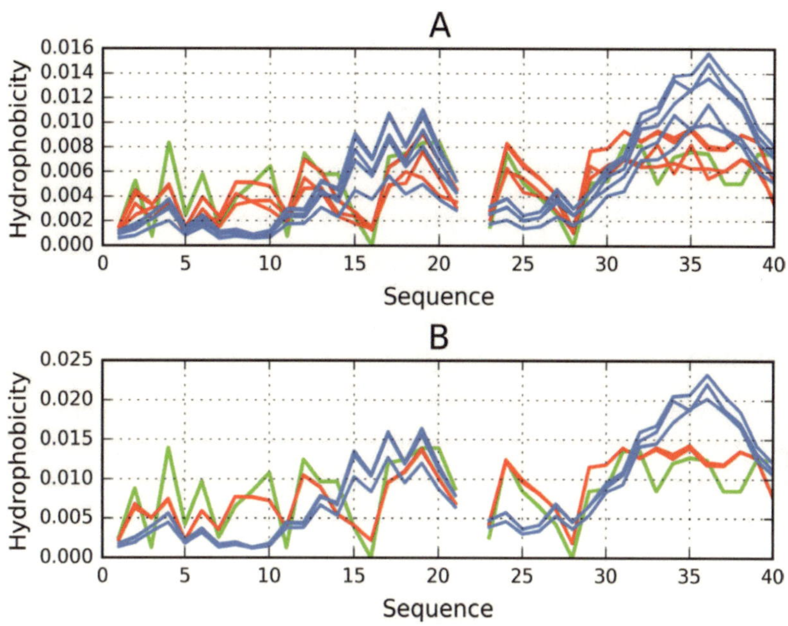

Figure 11. Status of protofibril (Chains ABCDE)—T(blue), O(red), H(green). A—all chains shown, B—central chains (BCD)—the border chains removed to show the almost identical distribution in central chains.

The status of the protofibril is described by the following parameters: RD (T-O-R): 0.639; RD (T-O-H): 0.659. This means that the involvement of intrinsic hydrophobicity in shaping the protofibril's structure is greater than in the case of the superfibril. Correlation coefficients are 0.280, 0.365, and 0.718 for HvT, OvT, and HvO, respectively. This further shows that the conformation of the protofibril is dominated by the intrinsic hydrophobicity of its residues.

3.4.3. Structure of an individual chain present in Aß(1-40) protofibril

The status of the C chain (i.e., the central chain in the protofibril) is visualized in **Figure 12**. The chart reveals strong dominance of intrinsic hydrophobicity (**Figure 12A**) with the consequent

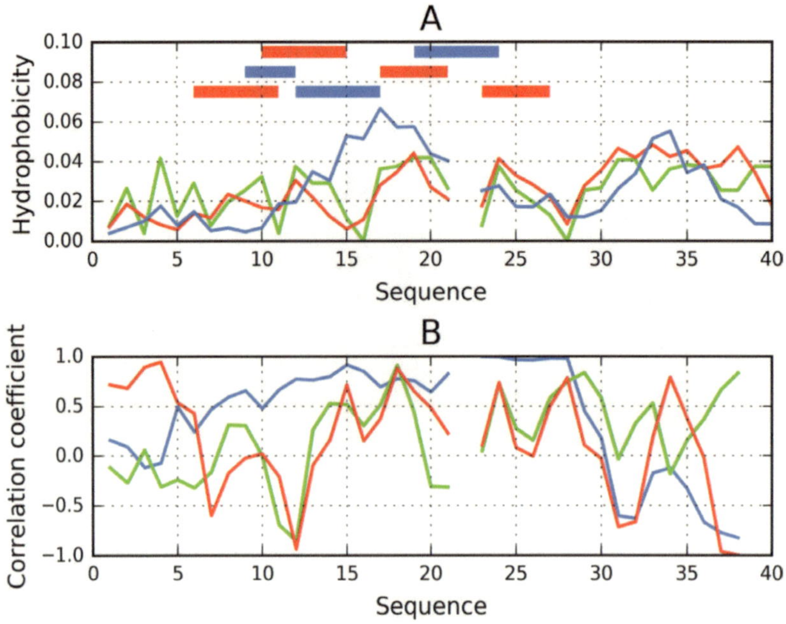

Figure 12. Aß(1-40) hydrophobicity profiles (2MVX). A—T(blue), O(red) and H(green) distributions. Horizontal lines correspond to local maxima (red) interspersed with local minima (blue). B—Correlation coefficients calculated for successive 5aa fragments to visualize the status of sequential 5 aa fragments (HvO—blue, TvO—red, HvT—green).

lack of alignment with the theoretical monocentric distribution (T). Comparing correlation coefficients for successive 5 aa fragments singles out the fragments at 9-16 and 22-27 as particularly discordant.

It should be noted that the charts plotted in **Figure 12A** and **B** for the central chain are also representative for all chains forming the protofibril, regardless of their length. The presence of local maxima and minima which are inconsistent with the theoretical (T) distribution applies to all chains. In this particular case, we can observe a local maximum at 7-9 (contrary to T, which predicts a local minimum), a local minimum at 9-11 (in place of the expected increase in hydrophobicity), then another maximum at 10-15, a minimum at 14-17 (where the T distribution predicts a global peak), a minimum at 22, and another local maximum at 22-27. The C-terminal fragment is consistent with the theoretical distribution; however, the calculated RD values and correlation coefficients confirm strong discordance vs. T (**Table 4**).

Figure 13 provides a visualization of the linear propagation of alternating bands of high and low hydrophobicity.

Such linear arrangement of alternating bands can be expected to have an impact upon the properties of the aqueous environment. Experimental research, which focuses on water in contact with hydrophobic surfaces, suggests that under such circumstances, levitation of water particles may occur [49]. This phenomenon provides water particles with greater mobility, which, in turn, disfavors the formation of ice crystals. Since similar conditions are encountered on the surface of amyloids and antifreeze proteins, we may speculate that water does indeed gain increased mobility when in contact with amyloids—perhaps even interfering with the action of proteolytic enzymes [50], which are known not to degrade amyloid

Fragment	RD		Correlation coeff.		
	T-O-R	T-O-H	HvT	TvO	HvO
1-40	0.649	0.686	0.310	0.322	0.779
7-9	0.509	0.459	0.812	0.589	0.795
9-11	**0.522**	**0.808**	**−0.227**	**0.135**	**0.934**
10-15	**0.776**	**0.815**	**−0.330**	**−0.566**	**0.855**
14-17	0.493	0.269	0.425	0.275	0.913
16-22	0.427	0.359	0.477	0.521	0.947
22-27	0.509	0.460	0.242	0.190	0.987
27-40	0.615	0.545	0.352	0.577	0.719

Fragments listed in boldface exhibit strong amyloid characteristics.

Table 4. Status of the C chain (the central one in protofibril) and its individual fragments—RD values (T-O-R; T-O-H) and correlation coefficients (HvT, TvO, and HvO).

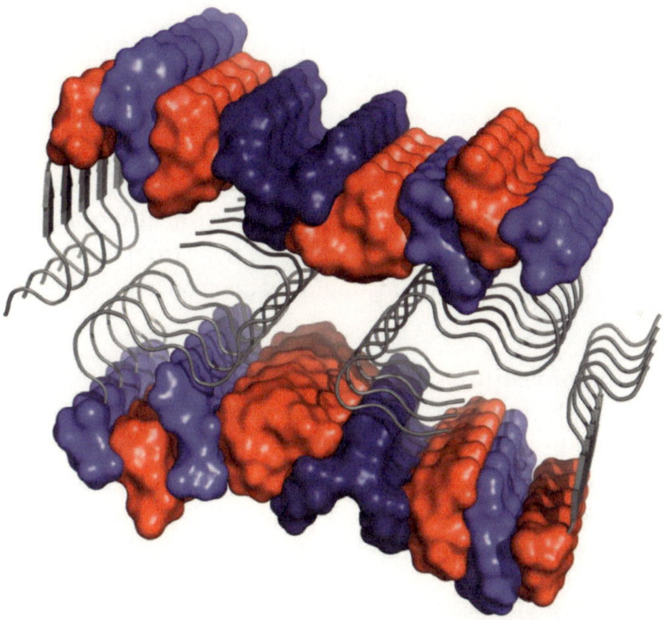

Figure 13. 3D presentation of 2MVX, highlighting local maxima (red) (8-9, 12-13, 17-19, and 23-24) and minima (blue) (10-11, 14-16, 20-21, and 25-26) (dark blue—fragment 14-16 with the minimum which is against the expected maximum—see **Figure 11A**) of hydrophobicity, as observed in Aß(1-40). The linear propagation of discordant fragments along the fibril's axis can be readily observed. It is possible to put here the arrows (as shown in **Figure 3**) to visualize the predicted variability in the structural properties of water (with orientation toward hydrophilic surfaces regarded as more likely), showing how the peculiar conditions occurring on the protein's surface disrupt the natural structuralization of the solvent.

fibrils. In effect, the enzyme fails to recognize the "signal" generated by an aberrant protein. Additionally, the amyloid protein is deprived of any fragments of structure accordant with the model as it is observed in antifreeze proteins, which act as native, biologically active proteins which undergo standard degradation procedure.

4. Conclusions

The analysis presented in this chapter, along with numerous other publications which deal with the fuzzy oil drop model, suggests that deviations from the theoretical distribution of hydrophobicity in a protein (i.e., exposure of hydrophobic and/or hydrophilic residues on the surface) are strongly linked with biological activity. This effect clearly disrupts the natural structuralization of the surrounding solvent and forces it to adapt itself to the protein's presence. In light of the presented analysis, the protein may be perceived as a more or less imperfect micelle (whether spherical, flattened, worm-like or cylindrical), devoid of the strong symmetries observed in surfactant micelles. Such imperfections (i.e., deviations from a perfectly symmetrical micelle) enable the protein to perform its intended function by interacting with other molecules (ligands or external proteins); however, such local deviations from optimal water-protein interactions also have an effect on the aqueous solvent itself. The degree to which the protein deviates from the ideal (theoretical) distribution of hydrophobicity is also a measure of the changes induced in its environment; such changes may be regarded as a means of communication and a carrier of information. This observation echoes the underlying assumptions of the so-called iceberg model [32].

The influence of water upon the native form of proteins remains an open issue—particularly in the context of local discordances, which are disadvantageous from the perspective of protein-solvent interactions. It is uncertain whether such discordances are only due to the nature of the residue sequence, which cannot produce a fully ordered micelle, or whether some other factors are at play. On the other hand, it is evident that such local disruptions affect the properties of the solvent (viewed as a continuum). It can also be demonstrated—on the grounds of the fuzzy oil drop model—that local deviations from the theoretical distribution of hydrophobicity are targeted and encoded information concerning the protein's biological activity [51].

Our approach likens the protein to an "intelligent micelle" which imposes a specific local disorder within the aqueous solvent and thereby encodes a description of the required protein-water interaction. A surfactant micelle—being perfectly symmetrical—carries no such information. In a similar vein, antifreeze proteins that are highly consistent with the theoretical distribution of hydrophobicity (such as the type II antifreeze protein [16]) perform their function simply by being present in the solvent and do not need to encode any specific information.

It appears that experimental research focusing on such "imperfect" micelles and on their interaction with the environment may yield clues regarding the function of proteins and, in particular, their specificity. Under these assumptions, the structure of the protein represents a very delicate balance between micelle-like ordering (which provides the protein with solubility) and local deviations, whose purpose is complex. On the one hand, such deviations encode the capacity to perform a specific chemical reaction as a result of inter-molecular contact; on the other hand, they represent signals which manifest themselves by the changes in the solvent as it adapts itself to the presence of the protein.

The solenoid fragment included in certain antifreeze proteins produces more or less parallel bands with very different structural properties in the surrounding medium. Whatever the ordering in the neighborhood of hydrophobic bands, it definitely does not resemble the structure observed near hydrophilic bands. Experimental studies indicate that water gains increased mobility at the boundary between bands [42, 49]. Such dynamic properties may prevent formation of ice crystals, which otherwise calls for highly uniform conditions in the bulk medium. Nevertheless, the antifreeze protein is not merely a solenoid—it also includes other fragments (with diverse secondary structural characteristics), mediating optimal interaction with the water environment. While similar linear arrangement can be observed in amyloids, they differ from antifreeze proteins in the sense that their ideal periodicity is dependent on the sequential identity of each complexed chain. We may speculate that under such conditions, the aberrant protein issues a signal which cannot be recognized by other proteins—this would explain why enzymes do not interact with the amyloid.

The concept of reorganization of the aqueous solvent may even be approached from a rheological perspective [52].

The mechanisms employed by antifreeze proteins resemble those exploited on a macroscopic scale by humans—for example, inducing mobility (e.g., by stirring, which prevents freezing) or by introducing factors which produce structural changes in the medium, disfavoring the formation of ice (such as salting). The "tasks" performed by ancillary fragments of the antifreeze protein are associated with the specific nature of the environment in which proteins operate—in particular, their role is to ensure solubility. A precipitated antifreeze protein would be useless; hence, the long helix runs parallel to the solenoid and ensures solubility by limiting the "disorder" which the solenoid fragment induces. The list of similarities between macroscopic processes and molecular-scale phenomena exploited by living organisms is long. A more detailed discussion of the subject can be found in [53].

The information encoded by a 3D structure concerns not only the specifics of a molecular process (lowering the energy threshold for enzymatic reactions) but also—or perhaps most importantly—the means by which different molecules may communicate with one another. Substrate recognition is not merely based on a "lock and key" mechanism. The participating molecules must first recognize each other, and this recognition may exploit distortions in the structure of the aqueous solvent. Under this hypothesis, the solvent acts as a carrier of information, which is specific enough to be recognized by the intended recipient. Significant research interest has recently been directed at biological systems in dried-out conditions (such as seeds [54]). When most of the solvent is eliminated from the system, leaving only enough water to preserve the 3D conformation of proteins (as in a dry seed), communication effectively ceases. However, this change is reversible, and communication can be restored simply by adding water—this causes the seed to resume its biological function and germinate.

Acknowledgements

The authors wish to thank Piotr Nowakowski and Anna Śmietańska for editorial assistance. This research has been supported by Jagiellonian University Collegium Medicum grant no. K/ZDS/006363.

Author details

Mateusz Banach[1], Leszek Konieczny[2] and Irena Roterman[1*]

*Address all correspondence to: irena.roterman-konieczna@uj.edu.pl

1 Department of Bioinformatics and Telemedicine, Jagiellonian University—Medical College, Kraków, Poland

2 Chair of Medical Biochemistry, Jagiellonian University – Medical College, Krakow, Poland

References

[1] Anfinsen CB. The formation and stabilization of protein structure. The Biochemical Journal. 1972;**128**(4):737-749

[2] Anfinsen CB. Principles that govern the folding of protein chains. Science. 1973;**181**(4096): 223-230

[3] Dobson CM. Protein folding and misfolding. Nature. 2003;**426**(6968):884-890

[4] Hartl FU. Molecular chaperones in cellular protein folding. Nature. 1996;**381**(6583): 571-579

[5] Dill KA, Ozkan SB, Shell MS, Weikl TR. The protein folding problem. Annual Review of Biophysics. 2008;**37**:289-316

[6] Muñoz V, Cerminara M. When fast is better: Protein folding fundamentals and mechanisms from ultrafast approaches. Biochemical Journal. 2016;**473**(17):2545-2559

[7] https://www.ebi.ac.uk/casp1/Casp1.html

[8] Berendsen HJ, Van Gunsteren WF, Zwinderman HR, Geurtsen RG. Simulations of proteins in water. Annals of the New York Academy of Sciences. 1986;**482**:269-286

[9] Van Gunsteren WF, Berendsen HJ. Molecular dynamics: Perspective for complex systems. Biochemical Society Transactions. 1982;**10**(5):301-305

[10] Van Der Spoel D, Lindahl E, Hess B, Groenhof G, Mark AE, Berendsen HJ. GROMACS: Fast, flexible, and free. Journal of Computational Chemistry. 2005;**26**(16):1701-1718

[11] Piwowar M, Banach M, Konieczny L, Roterman I. Hydrophobic core formation in protein complex of cathepsin. Journal of Biomolecular Structure & Dynamics. 2014;**32**(7): 1023-1032

[12] Banach M, Konieczny L, Roterman I. The fuzzy oil drop model, based on hydrophobicity density distribution, generalizes the influence of water environment on protein structure and function. Journal of Theoretical Biology. 2014;**359**:6-17

[13] Bryliński M, Prymula K, Jurkowski W, Kochańczyk M, Stawowczyk E, Konieczny L, Roterman I. Prediction of functional sites based on the fuzzy oil drop model. PLoS Computational Biology. 2007;**3**(5):e94

[14] Dygut J, Kalinowska B, Banach M, Piwowar M, Konieczny L, Roterman I. Structural interface forms and their involvement in stabilization of multidomain proteins or protein complexes. International Journal of Molecular Sciences. 2016;**17**(10):E1741

[15] Brylinski M, Kochanczyk M, Broniatowska E, Roterman I. Localization of ligand binding site in proteins identified in silico. Journal of Molecular Modeling. 2007;**13**(6-7):665-675

[16] Banach M, Konieczny L, Roterman I. Why do antifreeze proteins require a solenoid? Biochimie. 2018;**144**:74-84

[17] Roterman I, Banach M, Konieczny L. Application of the fuzzy oil drop model describes amyloid as a ribbonlike micelle. Entropy. 2017;**19**(4):167

[18] Roterman I, Banach M, Kalinowska B, Konieczny L. Influence of the aqueous environment on protein structure—A plausible hypothesis concerning the mechanism of amyloidogenesis. Entropy. 2016;**18**(10):351

[19] Kelley EG, Murphy RP, Seppala JE, Smart TP, Hann SD, Sullivan MO, Epps TH. III Size evolution of highly amphiphilic macromolecular solution assemblies via a distinct bimodal pathway. Nature Communications. 2014;**5**:3599

[20] Nisticò R, Scalarone D, Magnacca G. Sol-gel chemistry, templating and spin-coating deposition: A combined approach to control in a simple way the porosity of inorganic thin films/coatings. Microporous and Mesoporous Materials. 2017;**248**:18-29

[21] Huang J, Hang S, Feng Y, Li J, Yan H, He F, Wang G, Liu Y, Wang L. Rheological properties and application of wormlike micelles formed by sodium oleate/benzyltrimethyl ammonium bromide. Colloids and Surfaces A: Physicochemical and Engineering Aspects. 2016;**500**:222-229

[22] Kumar S, Awang MB, Abbas G, Kalwar SA. Wormlike micellar solution: Alternate of polymeric mobility control agent for chemical EOR. Journal of Applied Sciences. 2014;**14**:1023-1029

[23] Sarvi MN, Stevens GW, Gee ML, O'Connor AJ. The co-micelle/emulsion templating route to tailor nano-engineered hierarchically porous macrospheres. Microporous and Mesoporous Materials. 2012;**149**(1):101-105

[24] Liu N, He Q, Wang Y, Bu W. Stepwise self-assembly of a block copolymer–platinum(II) complex hybrid in solvents of variable quality: From worm-like micelles to free-standing sheets to vesicle-like nanostructures. Soft Matter. 2017;**13**:4791-4798

[25] Patel V, Ray D, Singh K, Abezgauz L, Marangoni G, Aswal VK, Bahadur P. 1-Hexanol triggered structural characterization of the worm-like micelle to vesicle transitions in cetyltrimethylammonium tosylate solutions. RSC Advances. 2015;**5**:87758-87768

[26] Konieczny L, Brylinski M, Roterman I. Gauss-function-based model of hydrophobicity density in proteins. In Silico Biology. 2006;**6**(1-2):15-22

[27] Liepinsh E, Otting G, Harding MM, Ward LG, Mackay JP, Haymet AD. Solution structure of a hydrophobic analogue of the winter flounder antifreeze protein. European Journal of Biochemistry. 2002;**269**(4):1259-1266

[28] Nishimiya Y, Kondo H, Takamichi M, Sugimoto H, Suzuki M, Miura A, Tsuda S. Crystal structure and mutational analysis of Ca^{2+}-independent type II antifreeze protein from longsnout poacher, *Brachyopsis rostratus*. Journal of Molecular Biology. 2008;**382**(3):734-746

[29] Cheng J, Hanada Y, Miura A, Tsuda S, Kondo H. Hydrophobic ice-binding sites confer hyperactivity of an antifreeze protein from a snow mold fungus. The Biochemical Journal. 2016;**473**(21):4011-4026

[30] Schütz AK, Vagt T, Huber M, Ovchinnikova OY, Cadalbert R, Wall J, Güntert P, Böckmann A, Glockshuber R, Meier BH. Atomic-resolution three-dimensional structure of amyloid β fibrils bearing the Osaka mutation. Angewandte Chemie (International Ed. in English). 2015;**54**(1):331-335

[31] Kalinowska B, Banach M, Konieczny L, Roterman I. Application of divergence entropy to characterize the structure of the hydrophobic core in DNA interacting proteins. Entropy. 2015;**17**(3):1477-1507

[32] Kauzmann W. Some factors in the interpretation of protein denaturation. Advances in Protein Chemistry. 1959;**14**:1-63

[33] Kyte J, Doolittle RF. A simple method for displaying the hydropathic character of a protein. Journal of Molecular Biology. 1982;**157**(1):105-132

[34] Levitt M. A simplified representation of protein conformations for rapid simulation of protein folding. Journal of Molecular Biology. 1976;**104**(1):59-107

[35] Kullback S, Leibler RA. On information and sufficiency. Annals of Mathematical Statistics. 1951;**22**(1):79-86

[36] Kalinowska B, Banach M, Wiśniowski Z, Konieczny L, Roterman I. Is the hydrophobic core a universal structural element in proteins? Journal of Molecular Modeling. 2017;**23**(7):205

[37] Ben-Naim A. Myths and verities in protein folding theories: From Frank and Evans iceberg-conjecture to explanation of the hydrophobic effect. The Journal of Chemical Physics. 2013;**139**(16):165105

[38] Bar Dolev M, Braslavsky I, Davies PL. Ice-binding proteins and their function. Annual Review of Biochemistry. 2016;**85**:515-542

[39] Walters KR Jr, Serianni AS, Sformo T, Barnes BM, Duman JG. A nonprotein thermal hysteresis-producing xylomannan antifreeze in the freeze-tolerant Alaskan beetle Upis ceramboides. Proceedings of the National Academy of Sciences of the United States of America. 2009;**106**(48):20210-20215

[40] Miskowiec A, Buck ZN, Hansen FY, Kaiser H, Taub H, Tyagi M, Diallo SO, Mamontov E, Herwig KW. On the structure and dynamics of water associated with single-supported zwitterionic and anionic membranes. Chemical Physics. 2017;**146**(12):125102

[41] Fletcher GL, Hew CL, Davies PL. Antifreeze proteins of teleost fishes. Annual Review of Physiology. 2001;**63**:359-390

[42] Modig K, Qvist J, Marshall CB, Davies PL, Halle B. High water mobility on the ice-binding surface of a hyperactive antifreeze protein. Physical Chemistry Chemical Physics. 2010;**12**(35):10189-10197

[43] Banach M, Kalinowska B, Konieczny L, Roterman I. Role of disulfide bonds in stabilizing the conformation of selected enzymes—An approach based on divergence entropy applied to the structure of hydrophobic core in proteins. Entropy. 2016;**18**(3):67

[44] CATH. http://www.cathdb.info/

[45] Dawson NL, Lewis TE, Das S, Lees JG, Lee D, Ashford P, Orengo CA, Sillitoe I. CATH: An expanded resource to predict protein function through structure and sequence. Nucleic Acids Research. 2017;**45**(D1):D289-D295

[46] de Beer TAP, Berka K, Thornton JM, Laskowski RA. PDBsum additions. Nucleic Acids Research. 2014;**42**:D292-D296

[47] http://www.ebi.ac.uk/

[48] Roterman I, Banach M, Konieczny L. Propagation of fibrillar structural forms in proteins stopped by naturally occurring short polypeptide chain fragments. Pharmaceuticals. 2017;**10**(4):89

[49] Schutzius TM, Jung S, Maitra T, Graeber G, Köhme M, Poulikakos D. Spontaneous droplet trampolining on rigid superhydrophobic surfaces. Nature. 2015;**527**(7576):82-85

[50] Saido T, Leissring MA. Proteolytic degradation of amyloid β-protein. Cold Spring Harbor Perspectives in Medicine. 2012;**2**(6):a006379

[51] Banach M, Konieczny L, Roterman I. Secondary and supersecondary structure of proteins in light of the structure of hydrophobic cores. In: Alexander K, editor. Methods in Molecular Biology Protein Supersecondary Structures. Vol. 2. Springer. (in press)

[52] Georgiev MT, Danov KD, Kralchevsky PA, Gurkov TD, Krusteva DP, Arnaudov LN, Stoyanov SD, Pelan EG. Rheology of particle/water/oil three-phase dispersions: Electrostatic vs. capillary bridge forces. Journal of Colloid and Interface Science. 2018; **513**:515-526

[53] Konieczny L, Roterman I, Spólnik P. Systems Biology – Functional Strategies of Living Organisms. New York, Heidelberg, Dordrecht, London: Springer; 2013

[54] Gordon ME, Payne PI. In vitro translation of the long-lived messenger ribonucleic acid of dry seeds. Planta. 1976;**130**:269-273

Chapter 5

Superhydrophobicity through Coatings Prepared by Chemical Methods

Sepehr Shadmani, Mehdi Khodaei,
Xiuyong Chen and Hua Li

Additional information is available at the end of the chapter

http://dx.doi.org/10.5772/intechopen.92626

Abstract

Superhydrophobic surfaces were first observed in nature like on a lotus leaf. The surfaces need to have hierarchical micro- and nanoscale roughness and low surface energy to achieve superhydrophobicity. Their unique behavior against water leads to various applications like corrosion resistance, oil-water separation, self-cleaning properties, anti-icing properties, drag reduction, and antibacterial properties. To investigate the wetting behavior of the coating, water contact angle, contact angle hysteresis and sliding angle must be measured. If WCA is higher than 150° and sliding angle and contact angle hysteresis are below 10°, then it can be concluded that the surface is superhydrophobic. Various fabrication methods including lithography, templating, chemical vapor deposition, layer-by-layer deposition, colloidal aggregation, and electrospinning and electrospraying especially wet chemical method are thoroughly studied. Among all fabrication methods, the wet chemical technique is one of the promising methods due to its low cost and capability of large-scale production and also the substrate shape and dimensions having a minimal effect on the process. Superhydrophobic coatings still lack sufficient mechanical endurance. Also, in all traditional superhydrophobic coatings, it is necessary to lower the surface energy by a low-energy polymeric material that does not have suitable bonding and stability in harsh environments.

Keywords: superhydrophobic, coatings, chemical synthesis, surface engineering, self-cleaning, antimicrobial

1. Definition of superhydrophobicity

1.1. Concept

Superhydrophobic properties were first observed in nature and on the surface of *Nelumbo nucifera* (lotus), butterfly wings, *Brassica oleracea, Colocasia esculenta,* etc. [1, 2]. The superhydrophobic properties appeared due to unique surface structure and low surface energy. A superhydrophobic surface repels water droplets and does not get wet in contact with water. In other words, surface behavior against water is evaluated by the water contact angle measurements which will be discussed later. On a superhydrophobic surface, WCA is higher than 150°, while for hydrophobic and hydrophilic surfaces, this value is, respectively, 90°–150° and below 90°. In **Figure 1**, some natural superhydrophobic surfaces are introduced.

Higher WCA values mean that a water droplet tends to maintain a spherical shape on the surface. On the other hand, lower WCA shows the tendency of a water droplet to spread on

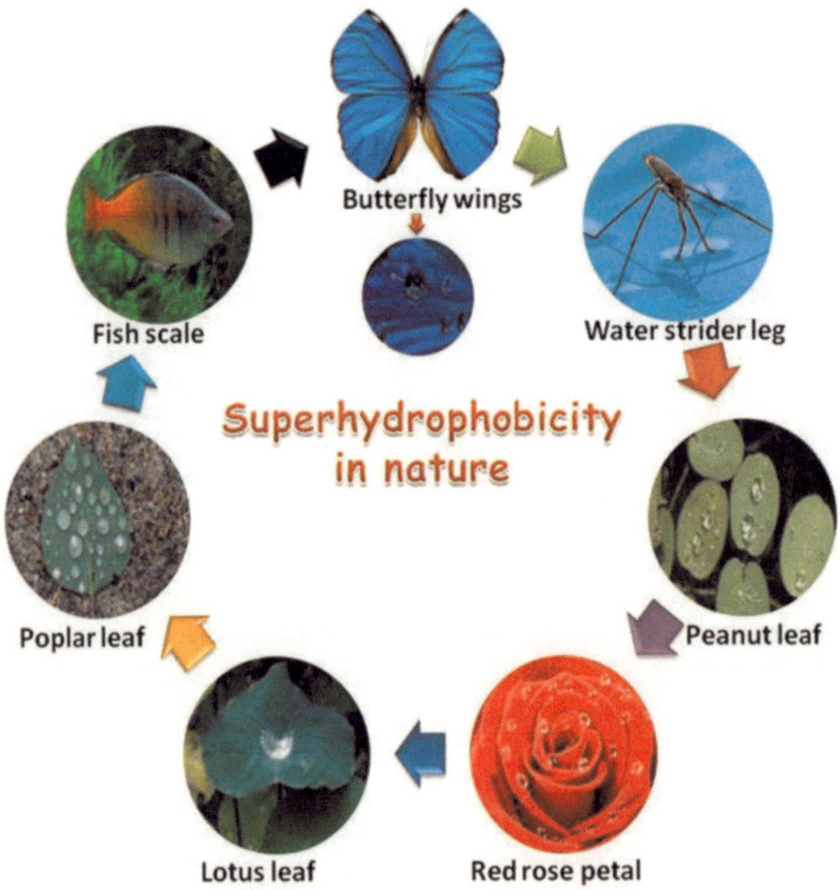

Figure 1. Superhydrophobicity in nature [3].

the surface. WCA is not the only parameter that is important to evaluate a superhydrophobic surface; other parameters like sliding angle and contact angle hysteresis are also important, which will show how slippery or sticky the surface is against a water droplet. These parameters will be discussed in depth later.

First, it is important to specify an ideal superhydrophobic coating. An ideal superhydrophobic coating has WCA higher than 150° (up to 180°, which is the theoretical limit), and also the sliding angle and contact angle hysteresis must be lower than 10° to guarantee low stickiness of the superhydrophobic surface against water.

This special wetting behavior will provide various special applications such as self-cleaning, anti-icing, antibacterial, oil-water separation, corrosion resistance, etc. for superhydrophobic surfaces and coatings.

1.2. Shortcoming and limitations

As mentioned before superhydrophobic surfaces and coatings have a special wetting behavior against water droplets which leads to various industrial applications. But one might ask: what hinders the application of these properties in industries?

Superhydrophobic coatings and surfaces are rather new due to their unique wetting behavior against water in comparison to other traditional coatings currently used in industries like powder and sol-gel coatings and other organic, inorganic, and metallic coatings. The traditional coatings do not possess high water contact angle and are usually hydrophilic, and more time is needed to improve quality and production costs of the superhydrophobic coatings.

The superhydrophobic coatings and surfaces must have two main features to achieve superhydrophobicity:

a. Hierarchical micro- and nanoscale roughness on the surface

b. Low surface energy

These two must be considered to fabricate a superhydrophobic surface. Various fabrication methods have been presented: These techniques are divided into two main categories including top-down and bottom-up. The top-down approach includes template-based techniques, lithography, and surface treatment by plasma. In the bottom to top approach, the structure is self-assembled and includes layer-by-layer deposition, chemical deposition, and colloidal assemblies. The methods to achieve superhydrophobicity are not limited to these methods, and there are several others like electrospinning, templating, chemical etching method, chemical vapor deposition, phase separation, electroless galvanic coating, sol-gel method, and thermal spray methods.

The main shortcoming of superhydrophobic coatings and surfaces is the low stability of the superhydrophobic properties or high cost of fabrication or lack of high-scale production capabilities. This leads to the limited use of superhydrophobic coatings. Although several promising approaches have been taken recently to increase mechanical stability, which will be discussed further in this chapter.

2. Surface wettability evaluation parameters

To investigate surface wettability, three main parameters are used. These three parameters are the water contact angle, contact angle hysteresis, and sliding angle. WCA is not enough alone to understand surface wetting behavior, and at least one (the CAH or SA) is needed to know how much water droplets stick to the surface. The definition of each parameter is provided below.

2.1. Water contact angle

Atoms and molecules of liquid and solid have higher energy on the surface because there are fewer chemical bonds on the surface. This energy is known as the surface tension or the surface free energy and shown by γ and is equal to energy per unit area needed to build surface in constant temperature and pressure (J/m^2 or N/m). In the case that solid and liquid are in direct contact with each other, the surface energy will be lower than in the situation in which these two are separated. The relation between surface energies and adhesion work is shown in the Dupre equation [14].

$$W_{SL} = \gamma_{SA} + \gamma_{LA} - \gamma_{SL}. \quad (1)$$

In this equation, W_{SL} is the adhesion work per unit area, γ_{SA} is the surface free energy between air and solid, γ_{LA} is the surface energy between air and liquid, and γ_{SL} is the surface free energy between liquid and solid.

When a water droplet is placed on the surface of the solid, these two will reach equilibrium, and the water droplet makes a specific angle with the surface known as water contact angle (θ_0). The below equation can calculate the total energy:

$$E_{total} = \gamma_{LA}\,(A_{LA} + A_{SL}) - W_{SL}A_{SL}. \quad (2)$$

In this equation, A_{LA} and A_{SL} are, respectively, liquid/air interface and liquid/solid interface. In this situation regardless of gravitational potential energy and in constant volume and pressure in the equilibrium, dE_{total} is considered equal to zero.

$$\gamma_{LA}\,(dA_{LA} + d\,A_{SL}) - W_{SL}d\,A_{SL} = 0. \quad (3)$$

For a droplet with constant volume, θ_0 can be calculated by the equation below:

$$dA_{LA}/dA_{SL} = \cos\,(\theta_0). \quad (4)$$

Then according to these equations, Cos θ_0 can be calculated by Young's equation.

$$\cos\theta_0 = (\gamma_{SA} - \gamma_{SL})/\gamma_{LA} \quad (5)$$

2.2. Contact angle hysteresis

To define contact angle hysteresis, first advancing (θ_a) and receding (θ_r) contact angles must be introduced. Contact angle hysteresis is calculated by subtraction of advancing and receding contact angles.

$$CAH = |(\theta||a - \theta_r)| \quad (6)$$

Consider a water droplet on the surface; if water droplets withdrawn or somehow evaporated from the surface, at first the surface area between the water droplet and surface does not change, but after a while, it starts to recede from the surface with a constant water contact angle equal to θ_r.

If at a controlled condition, the volume of water droplet increases by a syringe or is cooled down on the surface, at first, the volume increases without change in surface area in contact with the solid until it begins to advance on the surface with a constant water contact angle equal to θ_a.

Both advancing and receding contact angles on a surface depend on surface chemistry and topography, and a metastable droplet can have a contact angle between these two values which indicates the importance of measuring both of these values to evaluate surface wetting behavior [4].

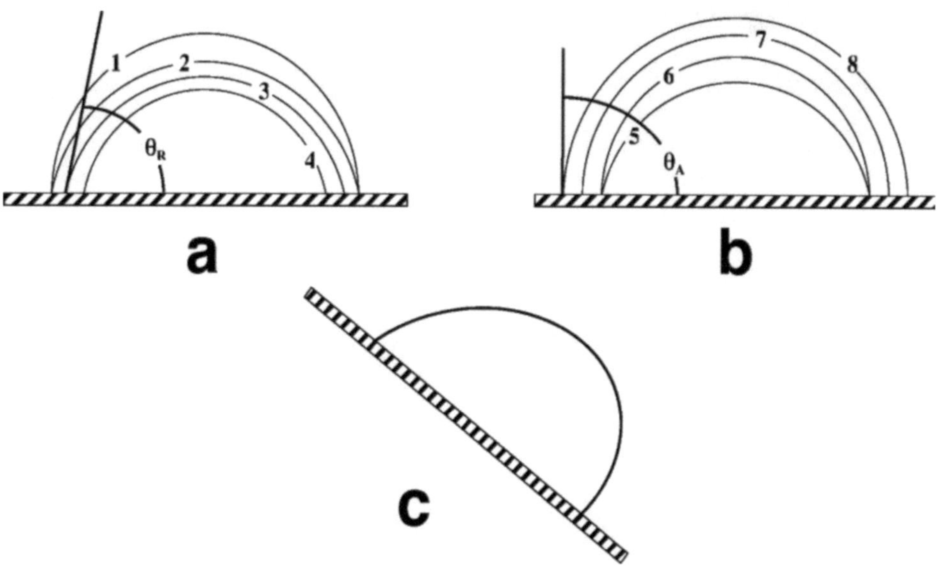

Figure 2. Schematics indicating the θ_a and θ_r (a and b, respectively) and a droplet on a tilted surface with θ_a in front and θ_r in the back [4].

Now consider a droplet on a tilted surface. While the droplet is moving downwards on a tilted surface, in the front, it expands, which will occur with a constant contact angle of θ_a, and on the back, it shrinks with a constant contact angle of θ_r which is shown in **Figure 2c** [4].

2.3. Sliding angle

The sliding angle is another parameter to evaluate the wetting behavior of the surface in which a droplet with a certain weight is dropped onto the surface and the sliding angle is the critical angle that a droplet starts to move and slide downwards. Sliding angle and contact angle hysteresis are both used to evaluate adhesion of droplet to surface. Contact angle hysteresis is more detailed and difficult to measure than the sliding angle [5].

3. Wetting models

3.1. Young's model

Several wetting models have been defined to calculate contact angle on the surface. The first wetting model is Young's equation that was just mentioned. This model does not consider surface roughness of the solid surface. Below Young's equation is shown.

$$cos\theta = \frac{\gamma_{SG} - \gamma_{SL}}{\gamma_{LG}} \tag{7}$$

In this equation θ is the contact angle, and γ_{SG}, γ_{SL}, and γ_{LG} are, respectively, the surface free energy of solid/gas, solid/liquid, and liquid/gas interface.

3.2. Wenzel model

It is obvious that in most cases the surface is not smooth, so Young's equation is not able to calculate the contact angle properly, so the Wenzel equation was introduced. In this equation, it is considered that the surface wetting occurs uniformly:

$$\cos\theta w = r\cos\theta \tag{8}$$

In this equation θw is the Wenzel contact angle, θ is Young's contact angle, and r represents the surface roughness factor that is equal to ratio of real surface to apparent surface.

3.3. Cassie-Baxter model

As mentioned before wetting is considered to be uniform in Wenzel's equation, or in other words, it is considered that water went through all surface cavities and there is no dry part. On the other hand, there is another wetting model which considers that the wetting is not uniform and air packets do not let water get into the surface cavities. In this case, water is in contact with solid and air packets, and water contact angle with air is equal to 180°. The model is called Cassie-Baxter and the equation is shown below:

http://dx.doi.org/10.5772/intechopen.92626

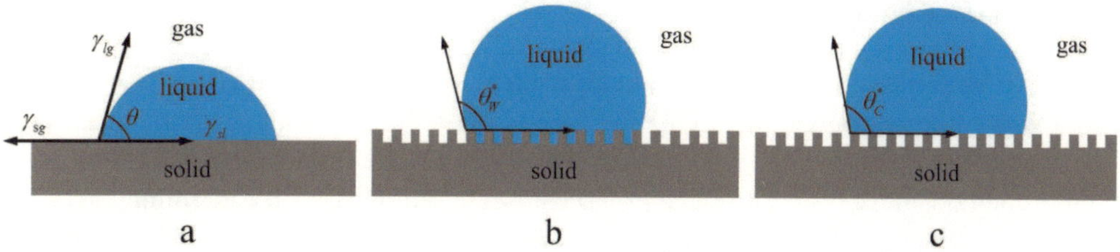

Figure 3. Schematics showing the difference between (a) young, (b) Wenzel, and (c) Cassie-Baxter wetting models.

$$\cos\theta_{CB} = f_1 * \cos\theta_1 + f_2 * \cos\theta_2 \quad (9)$$

$$\cos\theta_{CB} = f_1 * \cos\theta_0 + f_2 * \cos(\pi) \quad (10)$$

$$\cos\theta_{CB} = f_1 * \cos\theta - f_2 \quad (11)$$

$$\cos\theta_{CB} = f_1 * (\cos\theta + 1) - 1 \quad (12)$$

In the above equations, θ_{CB} is the Cassie-Baxter contact angle, f_1 is the ratio of the area that liquid is in contact with solid, and f_2 is the ratio of the area that liquid is in contact with air packets made or trapped inside the surface cavities. In **Figure 3** the difference between the three aforementioned wetting models is shown.

3.4. Transition between wetting models

In case of a hydrophobic surface or coating surface wettability respects to one of Wenzel or Cassie-Baxter models. In an ideal condition, a superhydrophobic coating should be seen in the Cassie-Baxter model. In the Cassie-Baxter model as mentioned before the topography and surface energy is in a way that droplet cannot penetrate through the empty space between micro- and nanoscale pillars on the surface while in Wenzel model the surface structure is large enough for water droplets to penetrate. Droplet adhesion to surface is more considerable in the Wenzel model than in the Cassie-Baxter model due to penetration of droplet into the micro- and nanoscale grooves on the surface.

Change in surface roughness and energy will lead to a transition from the Wenzel to the Cassie-Baxter model, which depends on the hierarchical micro- and nanoscale roughness on the surface. An easy way to evaluate whether the transition between Wenzel and Cassie-Baxter model has occurred or not is to measure the sliding angle. A noticeable decrease in sliding angle will be observed after the transition from the Wenzel to the Cassie-Baxter model due to increase in surface roughness and fabrication of hierarchical micro- and nanoscale roughness.

4. Applications of superhydrophobicity

Superhydrophobic surfaces and coatings as mentioned have a unique behavior against water droplets. This unique behavior results in a new set of applications including self-cleaning, anti-icing, antibacterial, oil-water separation, corrosion resistance, etc.

4.1. Oil-water separation

There have been many reports of oil contaminations in sea waters and rivers due to leak of factory waste into nature and accidents like Deep Water Horizon and Sanchi oil tanker collision. Removing oil contaminations from the water was always challenging and expensive, so different methods have been introduced by scientists to remove the oil contaminations. These methods are categorized into three main groups including water removal, oil removal, and smart controllable separators [6]. The water removing filters are superhydrophilic and superoleophobic; this kind of filter works underwater, and when they get wet by water, the presence of the water on the surface of the filter prevents oil to pass from the filter pores. The category in the oil removing method is a more efficient way because the amount of oil is always less than the amount of water, so it is logical that we try to remove the oil from water and not water from oil. To remove oil from water, the material should be superhydrophobic and superoleophilic; this mostly depends on the surface energy. The surface energy should be lower than the water surface tension (72.8 mN m^{-1}) and higher than the oil surface tension

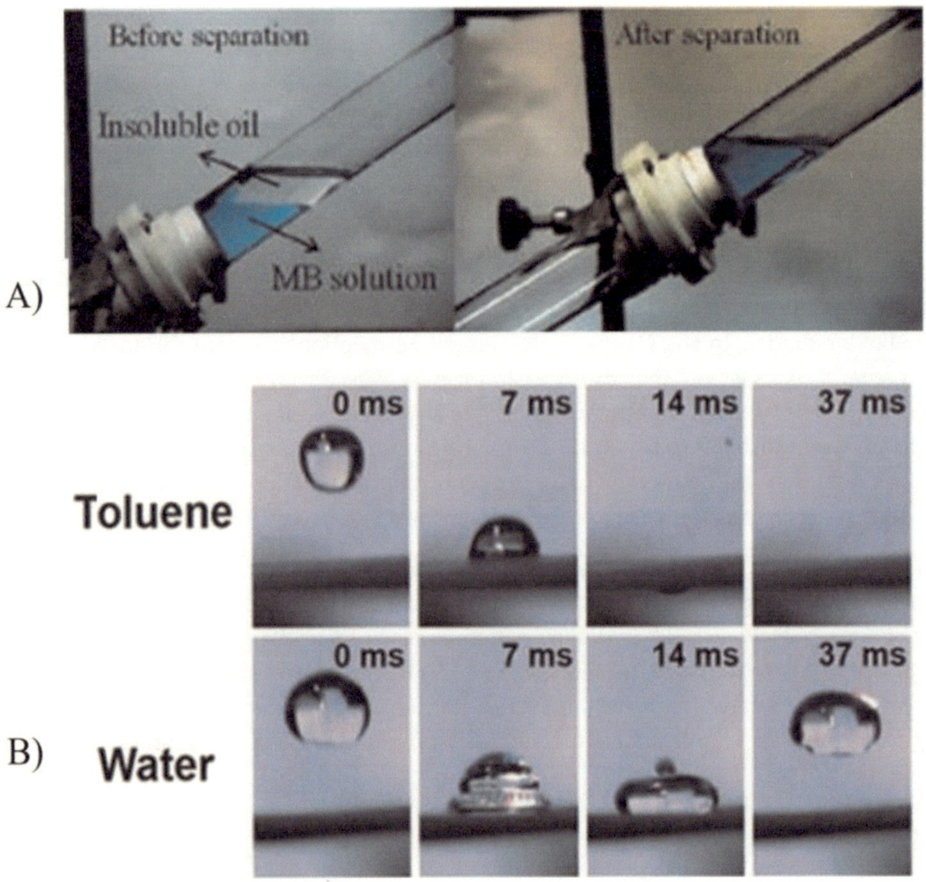

Figure 4. (A) Oil-water separation with the use of TiO_2-coated superhydrophobic and superoleophilic mesh, (B) opposite behavior of silicone elastomer-coated mesh against water and toluene droplets [7].

(30 mN m^{-1}). The oil removing method has two subcategories including oil removing filters and oil absorbents (like oil absorbent sponges, etc.)

Superhydrophobic oil removing filters are the main part of the oil removing category. Feng et al. [7] used a TiO_2-coated mesh to separate oil from water (see **Figure 4a**). Parkin et al. [8] used a silicone elastomer coating on a mesh to efficiently separate organic solvents like hexane, petroleum ether, and toluene from water. As shown in **Figure 4b**, the water droplet cannot pass through the filter, but toluene can easily pass through.

Also, absorbent materials are considered as a part of this group that can collect oil and changes it from liquid to a semi-solid phase. Tai et al. [9] built a graphene base sponge with high sensitivity and suitable recyclability (**Figure 5a**). The sponge was able to absorb oil up to 165 times of its weight. Pan et al. [10] built a three-dimensional superhydrophobic material through a one-step immersion process. This material had a high oil absorption capacity and was able to separate oil from water efficiently (**Figure 5b**). Superhydrophobic sponges could be used up to 300 times without losing their properties in an ideal situation. Currently, there are several serious challenges in this field. One of the main problems is the instability of the hierarchical structure of the coating on sponges that could easily get damaged by mechanical stresses or by exposure to chemical pollutions (acids, etc.). Also, most of the studies in this field worked on separation of oils with low density, and very few studies have been done on high-density oils [6].

4.2. Corrosion resistance

There are several ways to protect a surface from corrosion. One of the ways is to use different coatings or to use some processes to add heavy materials like chrome onto the surface which is harmful to the environment [11]. During the past two decades, scientists have been using superhydrophobic nanocomposite coatings without any toxic materials to protect various surfaces from corrosion [12–14]. The corrosion protection capability of the superhydrophobic coatings mainly is because of the presence of air pockets between surface and corrosive

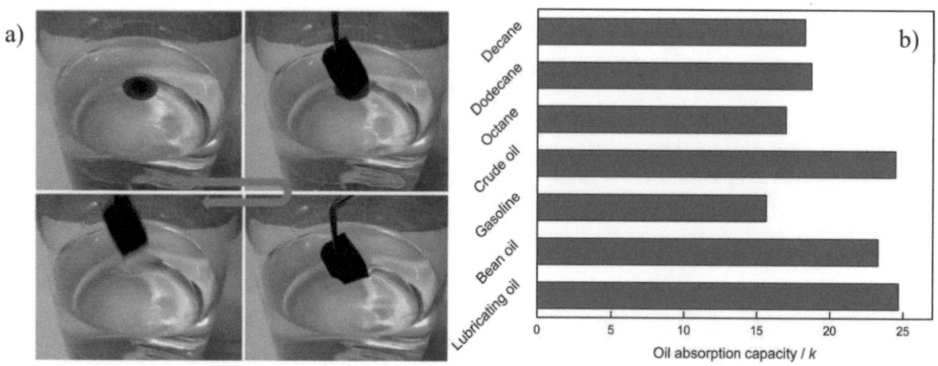

Figure 5. (a) Engine oil removing process using a superhydrophobic sponge [9], (b) the absorption capacity of superhydrophobic sponges for different nonpolar oil and solvents [10].

solution, and these packets act like a barrier and prevent corrosive ion diffusion and protect the substrate [15].

Advincula et al. [16] built a superhydrophobic corrosion-resistant nanostructure coating by using a conductive polymer in a two-step process. This coating could be deposited on any metal surface. They studied the corrosion resistance of the nanocomposite coating by the use of polarization test in NaCl solution in different pH and temperatures for 7 days, and the corrosion protection efficiency was reported to be higher than 95%.

Zhang et al. [17] worked on superhydrophobic membranes with different morphologies and chemical compositions through the sol-gel method. Humid air test and polarization tests showed insufficient corrosion protection. They realized that surface morphology is more important than the chemical composition of the sol-gel coating. In another study on the corrosion resistance of coatings on the Mg-Mn-Se alloy, three types of coatings with different wettabilities from hydrophilic to superhydrophobic were deposited on substrates, and corrosion resistance of the coatings in 3% NaCl solutions was studied [18]. Corrosion potential is known to be a criterion for corrosion resistance; the higher potential shows higher corrosion

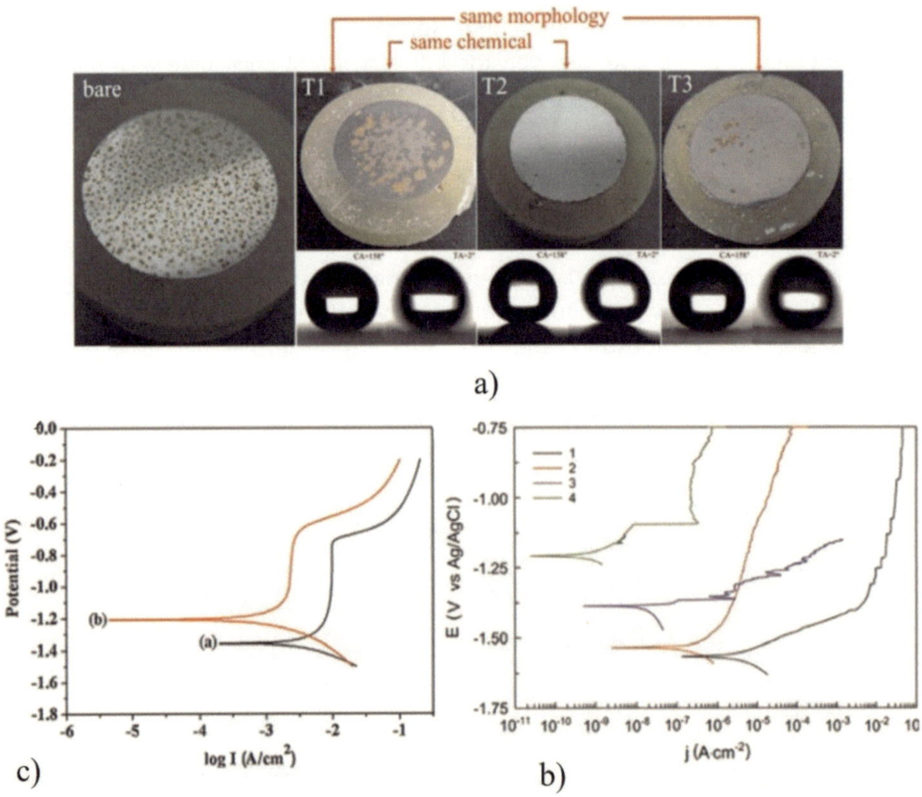

Figure 6. (a) Coating with different wettability after humid air test [17]; (b) polarization curve of the samples in 3.5%wt NaCl solution: (1) alloy without coating, (2) coated sample with PEO method, (3) hydrophobic coating, and (4) superhydrophobic nanocomposite coating [18]; and (c) polarization curve of the superhydrophilic Al (black line) and superhydrophobic PU/Al/Al_2O_3 in NaCl solution [20].

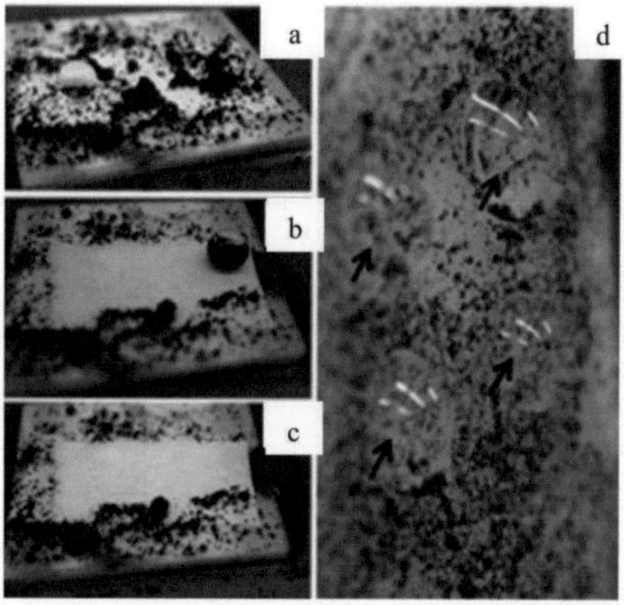

Figure 7. (a, b, c) Showing self-cleaning properties of the superhydrophobic high-density PE with TiO_2 nanoparticles; (d) adhesion of water droplet to high-density PE coating while the sample is held vertically [24].

resistance in general [19]. Superhydrophobic nanocomposite coating deposited with plasma electrolytic oxidation[1] on Mg alloy showed the best results in polarization tests (**Figure 6b**). Li et al. [20] managed to build a superhydrophobic corrosion-resistant polyurethane coating containing Al_2O_3 nanoparticles. The water contact angle of the coating with 2 wt% PU was 151°, and the sliding angle was 6.5°. An increase in corrosion potential showed the positive effect of superhydrophobic coating in corrosion protection (**Figure 6c**).

4.3. Self-cleaning properties

The lotus leaf's surface is always clean regardless of any contamination that may be present in its surrounding environment [21]. This leaf has a unique surface structure and is coated with wax and shows superhydrophobic properties, and the sliding angle is very low so water can easily slide on the surface of the leaf and remove any contamination. The aforementioned properties of superhydrophobic surfaces and coatings are called self-cleaning properties. Many superhydrophobic coatings were synthesized with different methods and used in industries, daily, or in military use [22, 23].

Lions et al. [24] produced a nanocomposite self-cleaning superhydrophobic high-density PE coating containing TiO_2 nanoparticles. Results showed that water droplets could remove big alumina particles or small graphite particles from the surface of the coating (**Figure 7a-c**). On the other hand, high-density PE by itself had a smooth surface which resulted in water

---

[1]REO.

droplets sticking on the surface of the coating, and the coating could not have self-cleaning properties (**Figure 7d**). Rodriguez et al. [24] also managed to build a coating based on lotus leaf surface morphology. This coating was made by nanostructure template assembly, and the sliding angle was between 4° and 7°. The self-cleaning properties of this coating were very close to that of a lotus leaf.

But the question is how a superhydrophobic surface has self-cleaning properties. The first reason is due to surface energy calculations. To explain how even hydrophobic particles can be collected by rolling drop will be further discussed.

When a spherical particle (pollution) is in contact with water on the sample surface (**Figure 8**), the area of the wetted surface can be calculated by the equation below in which 2Rs is the sphere diameter:

Area of the wetted part of the particle on the surface (pollution) = $2\pi R_s^2(1 + \cos\theta_{e.})$

Also, the liquid will lose some of the area of itself that can be calculated through the below equation:

Lost area of the liquid = $\pi R_s^2 * \sin\theta_e$

The change in surface energy can be calculated by the below equation:

$$\Delta F = 2*\pi*Rs^2*(1 + cos\theta e)(\gamma sl - \gamma sv) - \pi Rs^2 - sin\theta e*\gamma LV \tag{13}$$

$$\Delta F = \gamma LV*2*\pi*Rs^2*(1 + cos\theta e)^2 \tag{14}$$

When the equivalent water contact angle is not 0° and 180°, the particle always tends to attach to a spherical-shaped water droplet. The second reason for self-cleaning of these rough surfaces is that the contact area between the pollution particle and solid surface is very limited due to the unique surface roughness so the pollution particle has a very lower adhesion to the surface in comparison with smooth surface also water can diffuse into larger porosities as a result of impacting to the surface. The diffused water will absorb particles and get back to the top of the surface due to superhydrophobic properties and lead to self-cleaning properties [25].

Chen et al. [26] introduced a unique mechanism for self-cleaning surfaces by inspiration from Cicada wings (**Figure 9**). On this surface, pollutions are automatically removed due to the bouncing movement of water droplets on the surface. The ability of the pollutant particle to

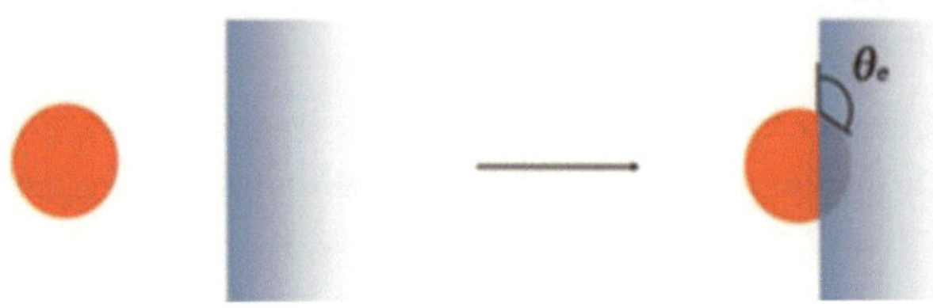

Figure 8. Schematic of the spherical particle that has moved from air into the water; the contact angle between the particle and water is shown [25].

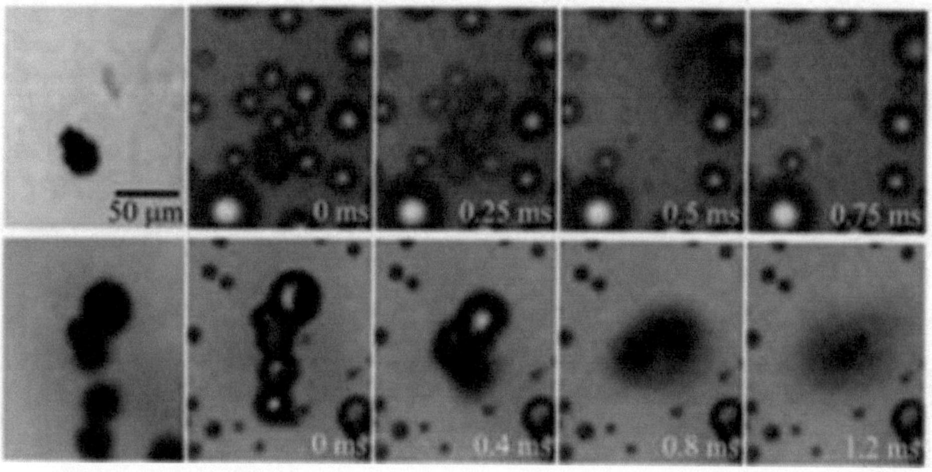

Figure 9. Self-cleaning properties of coating inspired from Cicada wing through bouncing movement of the pollutant particle on the surface [26].

bounce on the surface of the superhydrophobic coating mostly depends on the stability of the particle into the liquid phase. This unique coating shows there is a chance to produce and develop new self-cleaning coatings.

4.4. Anti-icing properties

Every year ice storms harm the electrical transmission equipment, communication systems, highways, etc. [27]. To reduce these kinds of damages, different methods of producing an anti-icing surface have been introduced [28]. Although there are other conventional methods like reducing icing temperature point and thermos electrical and mechanical methods, these methods use a lot of energy and are not economical.

In recent years superhydrophobic coatings have been suggested as an anti-icing coating. As mentioned before, the presence of air pockets on the superhydrophobic nanocomposite coating structure causes the water droplets to slide easily on the surface; therefore there will not be enough time for the droplet to frost on the surface [29–31]. In situations that the temperature is very low, superhydrophobic nanocomposite coatings can be used to prevent water from wetting the surface and cause frost and finally damage to the surface or equipment [32]. Chen et al. [33] deposited four types of coatings with different wettabilities from superhydrophilic to superhydrophobic on Al substrate. Dynamic studies of droplet impact to the superhydrophobic surface at low temperature showed that if the angle between the direction of droplet and surface of the coating is higher than 30°, then water droplet can easily slide and be removed from the surface. Hen et al. [34] produced a superhydrophobic nanocomposite film containing multi-walled silicon nanotubes. Results showed that the growing rate of ice on the non-coated Al surface is twice the surface with superhydrophobic nanocomposite coating (**Figure 10**).

In another study, an easy and low-cost nanocomposite coating containing polydimethylsiloxane with different coupling agents was investigated [35]. As shown in **Figure 11**, the superhydrophobic coating is completely effective in reducing ice adhesion to the surface up to 97%.

Scientists have some disagreements about the relations between superhydrophobicity and anti-icing properties. Some believe that these two are not related to each other; on the other hand, some insist that superhydrophobicity will result in anti-icing properties [36]. These disagreements are because there is no specific standard that can be used to evaluate ice adhesion to surfaces; also the method of preparing ice for each study is different from the

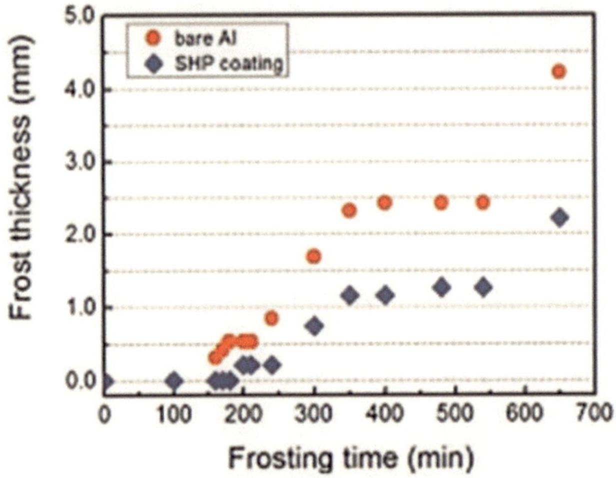

Figure 10. Comparison of ice growing rate on bare Al and Al with superhydrophobic nanocomposite coating containing multi-walled silicon nanotubes [34].

Figure 11. Comparison of the uncoated and superhydrophobic coated sample at −5°C and high humidity [35].

other, so by now it is not possible to have a definite answer to this matter [28]. The recent studies have helped to get a better understanding of the ice formation process on the superhydrophobic surface, but there is still much left unknown about the nucleation, growth, and adhesion to the surface which need more studies and information on this subject.

4.5. Drag reduction

One of the main problems that a solid moving in water like a submarine is facing is the drag force; this force has resulted from the friction force between water and solid surface which is moving through water. There are several examples in nature which show antidrag properties [37]. By inspiration from shark skin and lotus leaf, several superhydrophobic coatings were fabricated [38]. Here, the positive effect of superhydrophobicity on drag reduction will be discussed. As mentioned before superhydrophobic coatings have some air pockets inside their hierarchical micro- and nanoscale surface structure which will reduce the contact between solid and liquid so the drag force will dramatically reduce [39]. Drag reduction phenomenon by superhydrophobic surfaces was first reported in 1991 [40]. Muan et al. [41] studied the effect of superhydrophobic nanocomposite coating on drag reduction in linear and turbulent streams. This superhydrophobic coating contained TiO_2 nanoparticles and was deposited on

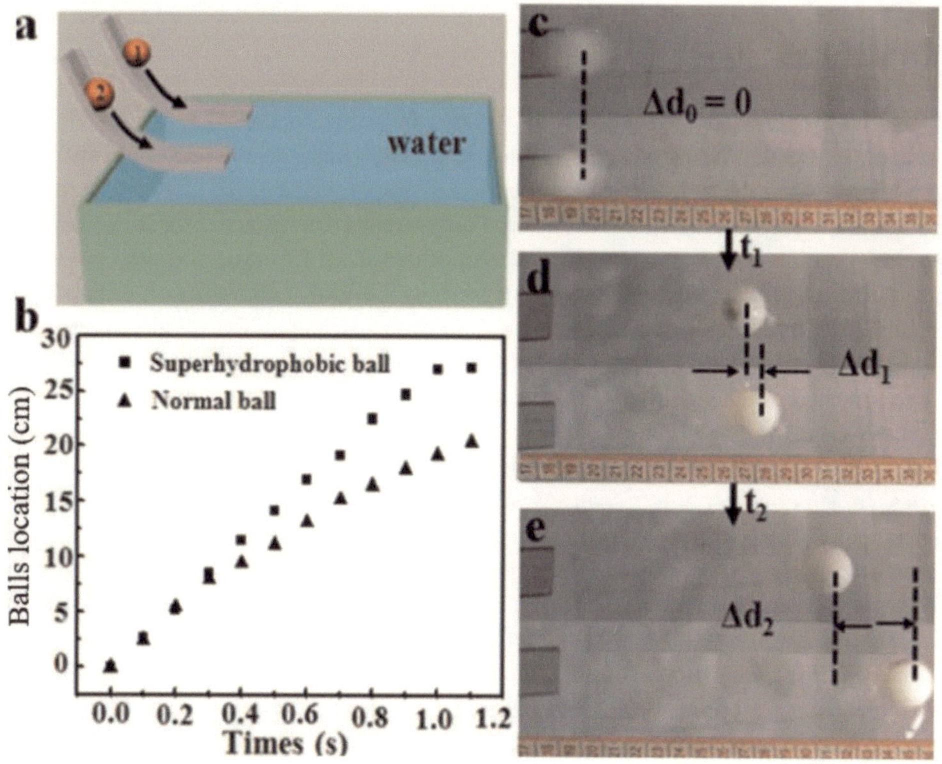

Figure 12. (a) Schematic of set up to throw balls into the water in the same condition, (b) balls' location-time diagram, (c, d, e) balls' picture at $t_1 = 0$ s, $t_2 = 0.61$ s, and $t_3 = 1.11$ s [42].

the Al substrate, and the drag force on this sample was compared with non-coated sample. Results showed that superhydrophobic coating will reduce drag force up to 30% for linear and 15% for a turbulent stream. Chen et al. [42] coated a ball with superhydrophobic coating; then they used a stage as shown in **Figure 12** to throw a ball with superhydrophobic coating and one without any coating; the ball with superhydrophobic coating had an average speed of 27.0 cm min^{-1}, but the average speed of ball without coating on water was 12.5 cm min^{-1}. This indicates that the superhydrophobicity is completely effective in the reduction of drag force and facilitates moving through water. On the other hand, Wei et al. [43] had a different opinion about this phenomenon. They believed that drag reduction is not because of the lower solid and liquid contact and the plastron effect is the main reason of this phenomenon. They fabricated a superhydrophobic coating by electrodeposition of gold on substrate. This superhydrophobic coating reduced the drag force up to 38.5% in speed of 0.46 m s^{-1} which is amazing. They said that reduction in water contact angle of the superhydrophobic coating will have a very small effect on drag reduction and will change it into 32.7%. They concluded that the main reason for drag reduction is not the high water contact angle but it is because of the plastron effect [44]. On a non-coated sample, the friction is just between solid and water, but on a superhydrophobic surface, there are three phases, water, solid, and trapped air between these two, so the friction will be drastically reduced in this situation that is known as the plastron effect.

4.6. Antibacterial properties

Antibacterial properties are essential in biosensors, implants, food packaging, and industrial and marine equipment [45, 46]. For example, one of the main reasons that cause infection in patients after surgery is bacteria that grow on implants [46]. To solve this problem, antibacterial coatings that reduce the bacterial adhesion to the surface or coatings containing antibacterial additives are suitable [47]. Schoenfisch et al. [47] produced a zero gel with the ability of nitrogen oxide release by spray method. In this case a combination of superhydrophobicity and nitrogen oxide release will result in a very strong antibacterial property. Nitrogen oxide showed a positive effect after some time and reduced the number of alive bacteria that had attached to the superhydrophobic surface. Ivanova and Philipchenko introduced an easy method to produce superhydrophobic coating by using chitosan nanoparticles. Antibacterial property is enhanced because of chitosan nanoparticles. Usage of nanosilver particles in superhydrophobic coatings also enhances the antibacterial properties; this enhancement is due to diffusion into the bacterial cell and damages the DNA structure from the inside [48]. There are still some doubts and questions about the mechanism of silver antibacterial properties. Heinonen et al. [49] fabricated a superhydrophobic coating containing silver nanoparticles through the sol–gel method. First silver nanoparticles were attached to γ-alumina by the Tollens process, and then the composite coating was functionalized with flouroalkyl silane[2] to reduce the surface energy. The diagram in **Figure 13** shows the number of bacteria on non-coated steel, superhydrophobic coated steel, and superhydrophobic coating

[2]FAS.

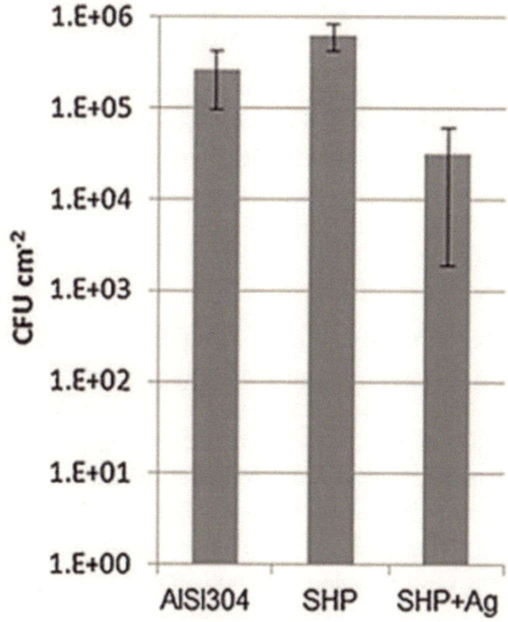

Figure 13. It shows the number of bacteria on non-coated steel (AISI304), superhydrophobic coated steel (SHP), and superhydrophobic coating with nanosilver particles steel (SHP + Ag) after 1-day exposure to bacteria at 25°C [49].

with nanosilver particle steel after 1-day exposure to bacteria at 25°C; the superhydrophobic coating with silver nanoparticles reduced the alive bacteria on the surface up to 88%.

Xue et al. [50] used the same method to fabricate silver nanoparticles on cotton fibers. After that hexadecyltrimethoxysilane was used to modify these fibers, and superhydrophobicity was achieved. As shown in **Figure 14**, non-coated cotton fiber does not have any resistance

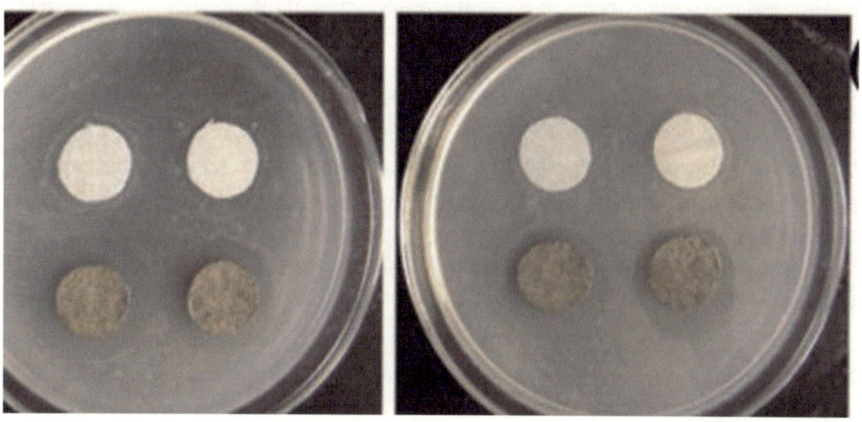

Figure 14. Comparison of antibacterial properties on non-coated cotton fiber (the white ones) and superhydrophobic coated fiber cottons (the gray ones) [50].

against bacteria, but on the other hand, the superhydrophobic coated sample with nanosilver additives destroys almost all of the bacteria from its surface.

In general, superhydrophobic coatings are not an ideal antibacterial coating, and in some cases, instability, toxicity and low durability of these coatings make them a problematic method for antibacterial purposes. So, further studies in this field are needed to overcome the current problems.

5. Fabrication methods

Several techniques have been introduced to fabricate superhydrophobic surfaces. These techniques are divided into two main categories including top-down and bottom-up. The top-down approach includes template-based techniques, lithography, and surface treatment by plasma. In the bottom to top approach, the structure is self-assembled and includes layer-by-layer deposition, chemical deposition, and colloidal assemblies. The methods to achieve superhydrophobicity are not limited to these methods, and there are several others like electrospinning, templating, chemical etching method, chemical vapor deposition, phase separation, electroless galvanic coating, sol-gel method, and thermal spray methods.

5.1. Wet chemical

5.1.1. Chemical etching method

Chemical etching and other methods like plasma etching can be used to introduce micro- and nanoscale roughness to the substrate. Also, this process can be combined with other methods of fabrication superhydrophobic coatings to improve special surface roughness that is needed. Almost in all studies, chemical etching and similar methods like that are followed by surface modification with low surface energy materials like fatty acids, fluoroalkyl silanes, etc. This kind of superhydrophobic surfaces only has laboratory usage. Because of their low stability, they are mainly used to study the behavior of superhydrophobic surfaces against water in different situations, but the combination of this method with others seems promising.

Song et al. [51] used $CuCl_2$ solution to do chemical substitution reaction on Al substrate to make hierarchical micro- and nanoscale roughness on the Al substrate which is necessary to achieve superhydrophobicity. After the chemical etching, the surface was modified with a fluoroalkyl silane solution to increase the WCA to higher than 150°. In this study, 1 molar $CuCl_2$ solution was used and different etching times were tried. Also, the effect of removing the deposited Cu on substrate was investigated. As mentioned before this method is a chemical substitution reaction and a Cu element deposit on the Al surface while Al^{3+} gets into solution; the deposited Cu on the surface does not have a chemical bond with the surface and can be removed by an ultrasonic bath in water. So, to study the effect of Cu removal from the substrate, samples were washed in an ultrasonic bath. After the chemical etching, all samples were put into a fluoroalkyl silane solution.

When the aluminum plate is immersed into the $CuCl_2$ solution, the chemical substitution starts, and copper ions react with the aluminum element on the surface, and aluminum chloride will be made by this reaction. As a result, the copper element will deposit on the surface. The aluminum corrosion potential is lower than copper, so when copper deposits on the surface of the aluminum, a galvanic reaction will occur, and reaction speed will be increased. This reaction is exothermic and produces a lot of heat. In addition to this, the copper on the surface reacts with solution water, and hydrogen ions will be produced which will make the solution acidic and be able to remove aluminum from the substrate. When the

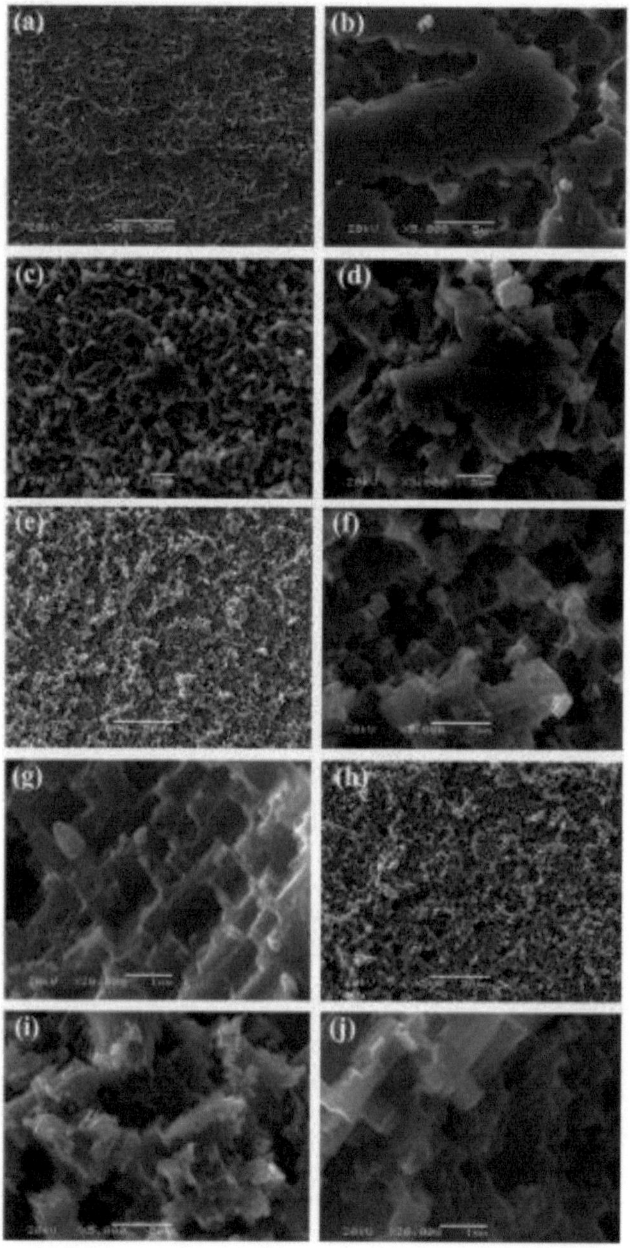

Figure 15. SEM image of etched aluminum surface in $CuCl_2$ solution in different periods of time: (a, b) 1 s, (c, d) 3 s, (e, f, g) 10 s, (h, i, j) 70 s [51].

hydrogen ions react with the aluminum on the surface of the sample, then a small hydrogen bubble will be made on the surface so during that time the copper ions cannot affect that part of the sample. As a result, the corrosion will not be uniform. This will be beneficial to achieve hierarchical micro- and nanoscale roughness. Below the reactions are mentioned.

$$2Al + 3Cu^{2+} \rightarrow 2Al^{3+} + 3Cu \tag{15}$$

$$Cu^{2+} + 2H_2O \leftrightarrow Cu\,(OH)_2 + 2H^{+} \tag{16}$$

$$2Al + 6H^{+} \rightarrow 2Al^{3+} + 3H_2\uparrow \tag{17}$$

The aluminum plate is a polycrystalline metal that has grain boundaries and dislocations; these places are ideal for corrosion and chemical substitution, so immediately after immersion of Al plate into the $CuCl_2$ solution, the reaction will take place in these places. As a result, rectangular planes and nanoscale steps will be formed on the surface. In **Figure 15**, SEM image of the surface after chemical etching for different periods is shown. It is worth mentioning that the samples were washed in water by ultrasonic bath [51].

- Efficient conditions to make a hierarchical structure on the Al substrate through chemical etching

In the aforementioned study, the effect of removal of deposited copper on the substrate was studied. In the results as can be seen in **Figure 16** in which the copper had been removed from the surface through ultrasonic bath, the stability of the superhydrophobicity is higher [51].

- Crucial role of surface modification after the chemical etching process

Chemical modification and reducing surface energy are necessary to achieve superhydrophobicity. Surface modification of the smooth surface with fluoroalkyl silane will increase the

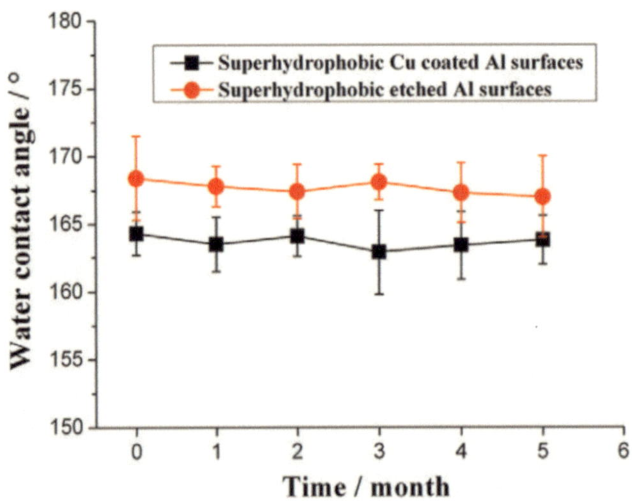

Figure 16. WCA measurement of the superhydrophobic samples during time (black line (with deposited copper), red line (without the deposited copper)) [51].

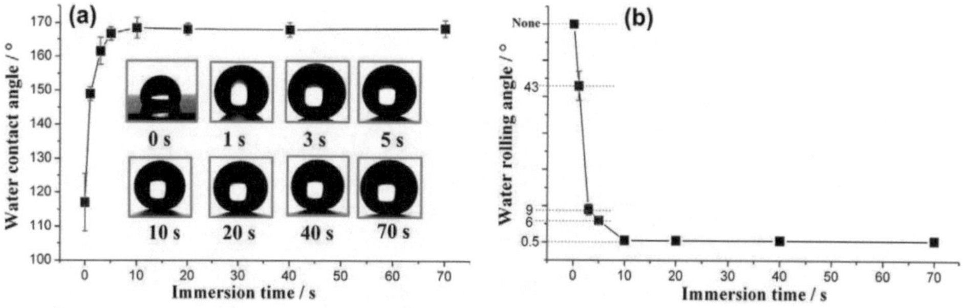

Figure 17. Change in WCA for different times of etching process (all samples were chemically modified with fluoroalkyl silane solution) [51].

WCA up to 117 which indicates the effect of surface energy. **Figure 17** shows WCA measurement results for samples with a different etching time which were chemically modified with fluoroalkyl silane solution before WCA measurements. As can be seen for etching time higher than 10 seconds, an enormous improvement in hydrophobicity is shown, and the superhydrophobicity was achieved. The reason for this improvement is the combination of hierarchical micro- and nanoscale roughness caused by chemical etching followed by low surface energy due to chemical modification with fluoroalkyl silane solution [51].

5.1.2. Sol-gel method

This method has been used in various studies to fabricate superhydrophobic coatings on different substrates. The sol-gel coating can be applied on the surface by various methods like dip coating, spin coating, spraying, etc. To achieve superhydrophobicity, the sol-gel coating is modified with a low surface energy material. Different approaches were taken to build a superhydrophobic coating, and some of them are mentioned here. After that, a more detail study about the sol-gel process will be provided.

One of the sol-gel coatings is alumina sol in which the aluminum tri-sec-butoxide was used as a precursor, and the particle size in sol was about 80 nm. The sol was deposited on the glass substrate by spin coating and cured at 400°C. The presence of nanoalumina particles in coating makes peaks on the surface of the coating with a height of $1\mu m$; and to achieve nano roughness, samples were put into boiling water for 5 minutes. As a result, a flower-like morphology was made on the coating, and after surface modification with low energy material, superhydrophobicity was achieved (look at **Figure 18a**) [52].

Mahadik et al. [53] used methyltrimethoxysilane as a precursor and fabricated a coating by the sol–gel method. The coating was deposited on the glass slides by dip coating and then cured at 150°C. After surface modification with trimethylchlorosilane, superhydrophobic properties were achieved. Coating WCA was 170 and it was stable up to 550°C but at temperatures higher than 600°C and after 2 h of exposure in this condition, the sample showed superhydrophilic properties which indicates that the surface chemical modification was destroyed. The superhydrophobicity was restored after doing the surface chemical modification with trimethylchlorosilane.

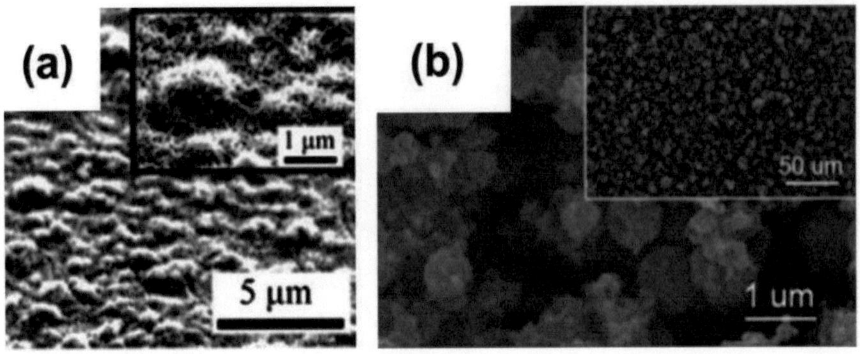

Figure 18. (a) SEM image of nanoalumina coating surface after 5-minute immersion into boiling water, (b) FESEM image of modified SiO_2 coating, which was deposited by electrospray method [53, 54].

Kim et al. [54] fabricated a superhydrophobic silica coating by electrospraying of sol–gel solution on the substrate. Tetraethoxysilane and methyltriacetoxysilane were used as sol–gel precursors. To achieve superhydrophobicity and self-cleaning properties, the coating was modified by perfluoro-octyl silane (**Figure 18b**).

Various studies have been reported in which sol-gel method was used to fabricate a rough surface and the low surface energy materials were used to reduce the surface energy and reach superhydrophobic properties, but almost all of the coatings suffered from lack of superhydrophobic property stability [55].

Superhydrophobic properties have been achieved on aluminum and silicon substrates by first fabrication of a rough surface on the substrate by either chemical bath deposition or electrochemical deposition or chemical etching, and then low surface energy treatments were done by FAS-17, stearic acid, or rf-sputtering of PTFE films [56–61].

Brassard et al. fabricated a superhydrophobic coating using the sol–gel method. They used the Stober process to fabricate SiO_2 nanoparticles from TEOS precursor, and then the synthesized nanoparticles were functionalized by fluoroalkyl silane. The functionalized SiO_2 nanoparticles were spin-coated in Al6061 substrate and dried at 70°C. The FTIR spectra results related to the functionalized SiO_2 nanoparticles are shown in **Figure 19**. The peaks 610 cm^{-1}, 730 cm^{-1}, 960 cm^{-1}, 1000 cm^{-1}, and 1250 cm^{-1} are related to C—F bonds in CF, CF_2, or CF_3 [62–65]. Also, the peaks at 1145 cm^{-1} approve chemical bonding between SiO_2 particles and fluoroalkyl silane. The peaks at 430 cm^{-1}, 800 cm^{-1}, and 1100 cm^{-1} are related to Si—O—Si bond.

The main problem with almost all the aforementioned superhydrophobic coatings is their lack of mechanical properties due to low adhesion and cohesion of the coatings. The articles about these kinds of superhydrophobic coatings do not consider the mechanical properties and stability, and they just focus on the effect of other coating parameters on the WCA of the coating. Khodaei and Shadmani [66] fabricated a superhydrophobic nanocomposite coating using the sol-gel method. The substrate was commercially available AA1050. The substrate

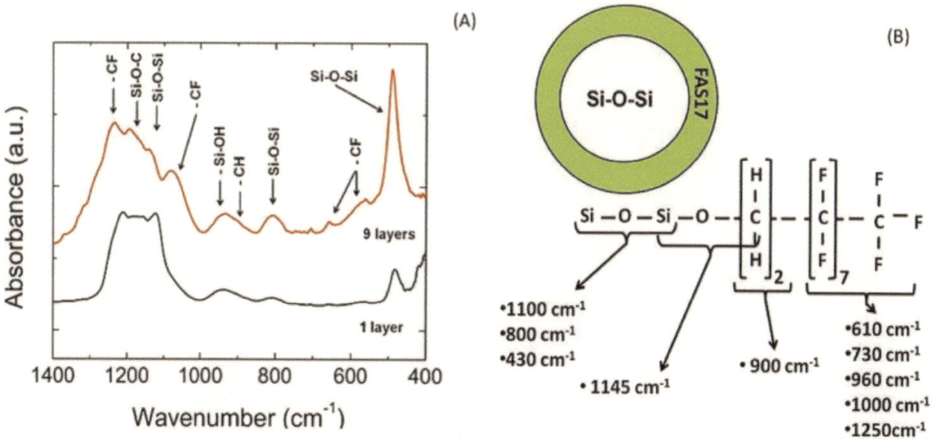

Figure 19. FTIR spectra of fluorinated silica nanoparticles coated on Al substrate as (A) a function of the number of layers and (B) schematic of the functionalized silica nanoparticle [55].

was first sanded and washed by acetone and then chemically etched to achieve micro roughness on the surface. Several approaches were compared to observe the effect of TEOS and GPTMS hybrid sol-gel coating containing functionalized Al_2O_3 nanoparticles. They used functionalized nano-Al_2O_3 particles to improve surface micro- and nanoscale roughness and also improve mechanical properties of the superhydrophobic coating. In **Figure 20** the manufacturing process and WCA measurements are reported. In total four samples were compared to each other: (a) chemically etched and then functionalized in FAS solution without sol-gel coating, (b) chemically etched substrate dip coated by sol-gel hybrid coating and then functionalized by FAS solution, (c) addition of Al_2O_3 to the hybrid sol-gel coating, and (d) addition of functionalized Al_2O_3 to the sol-gel coating. Results showed that functionalized nanoparticles had a uniform dispersion in coating and fabricated a uniform hierarchical micro- and nanoscale roughness which is ideal for superhydrophobicity and also acted as a shield during abrasion cycles and protected from the surface and superhydrophobic properties after 200 abrasion cycles with a total length of 300 cm.

5.1.3. *Electroless galvanic deposition*

In this method, galvanic reactions are used to fabricate superhydrophobic coatings. The reaction starts with contact of metal ion with the surface of a metal with lower corrosion potential. The reaction would be spontaneous, so it is a low-cost and efficient method to make roughness on the surface of the metal. After this process, a low energy material is used to decrease the surface energy and achieve superhydrophobicity [67].

Xu et al. [68] fabricated a superhydrophobic feather-like silver coating on a copper substrate using electroless galvanic deposition method. The WCA was 169 and the sliding angle was 2. In **Figure 21**, the feather-like morphology of the silver coating is shown.

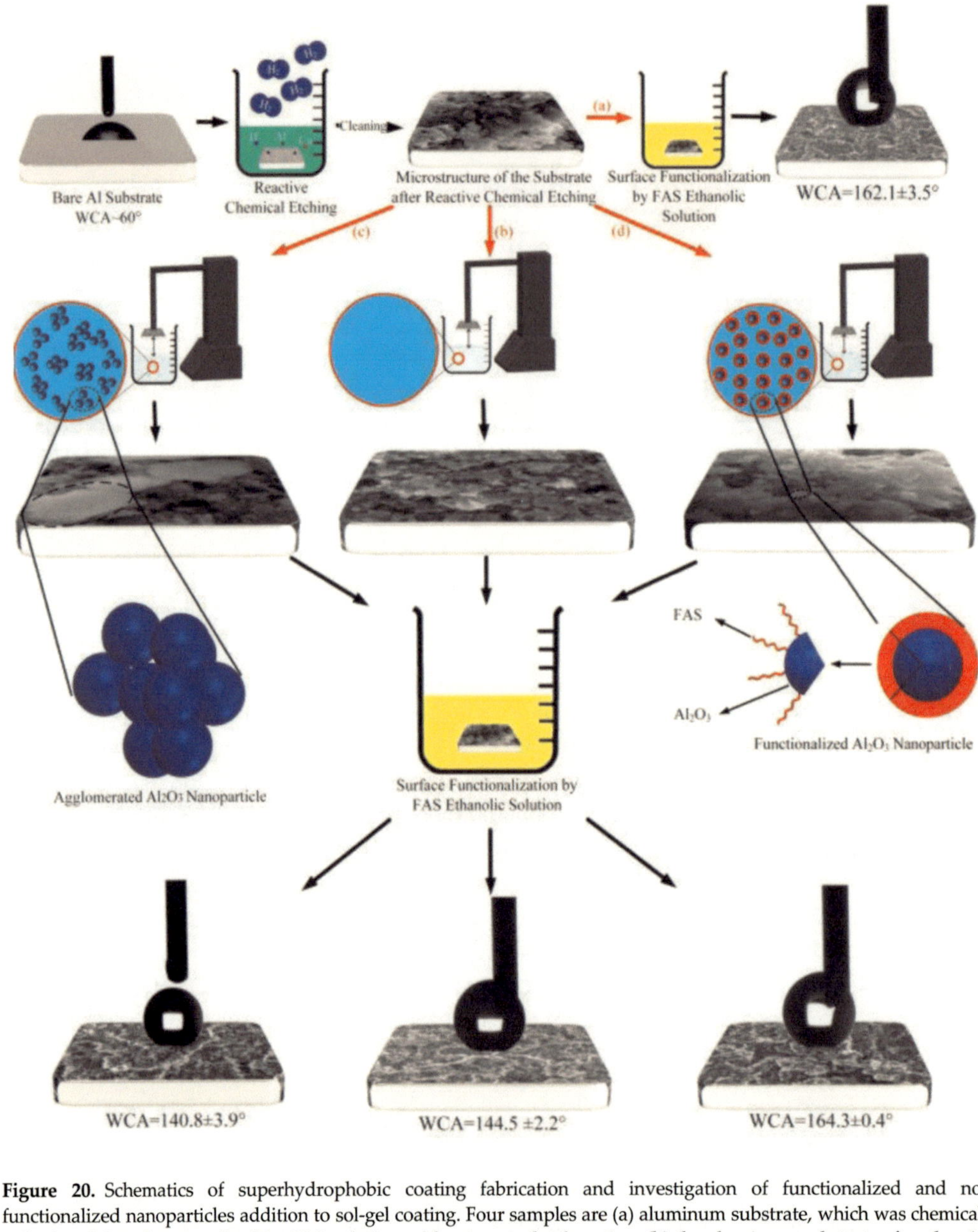

Figure 20. Schematics of superhydrophobic coating fabrication and investigation of functionalized and non-functionalized nanoparticles addition to sol-gel coating. Four samples are (a) aluminum substrate, which was chemically etched and then functionalized by FAS solution without any sol-gel coating, (b) the aluminum substrate after chemical etching was dip coated in TEOS-GPTMS hybrid sol-gel coating and the functionalized by FAS solution, (c) the same process was done by a hybrid TEOS-GPTMS coating containing Al_2O_3 nanoparticles and (d) functionalized Al_2O_3 was added to the coating and water contact measurements are reported [66].

5.1.4. Electrodeposition

The electrodeposition method is one of the chemical-based methods used to achieve superhydrophobicity. The main advantages of this method are its low cost of production,

Figure 21. SEM image showing the feather-like morphology of superhydrophobic silver coating on copper substrate [68].

capability of large-scale production, being independent from the shape and size of the sample, and great uniformity. Although it is worth mentioning that the electrodeposition technique fabricates a hierarchical micro- and nanoscale roughness on the surface of metal and a low energy material is needed to be coated on the surface after electrodeposition to reduce surface energy. A combination of the hierarchical structure on the surface and lowered surface energy by low energy material coatings like lauric acid, stearic acid, fluoropolymers, etc. will lead to superhydrophobicity. Also, several studies have been reported that used a two-step electrodeposition technique to fabricate ideal hierarchical micro- and nanoscale roughness on the surface [69, 70].

Jain et al. [71] fabricated a superhydrophobic copper substrate by electrodeposition technique followed by low surface energy modifications in stearic acid. The WCA was $162° \pm 3°$ and the sliding angle was about 3°. The copper substrate was chosen due to its wide application in industries. In **Figure 22**, SEM images of surface morphology and WCA are shown at different values of voltages including 0.5 V, 0.7 V, 0.9 V, and 1.1 V, showing the formation of globular asperities on the surface at potentials over 0.7 V.

The superhydrophobic surface fabricated by electrodeposition followed by stearic acid coating also showed self-cleaning properties. In **Figure 23**, superhydrophobic and ordinary surface are compared against an SiC particle dirt. As seen in the figure, the superhydrophobic sample cleaned completely by 55 drops of water.

In another study Xiang et al. [72] fabricated a superhydrophobic and superoleophilic mesh for oil-water separation using electrodeposition technique. The stainless-steel mesh was fist washed and degreased by ethanol and then etched by 8 M HCl to remove the oxidation layer from the surface. The prepared mesh then put into solution consists of ethanol, $CuSO_4$, Na_2SO_4, $NiSO_4$, NDM, and dopamine hydrochloride, and the electrodeposition was done at

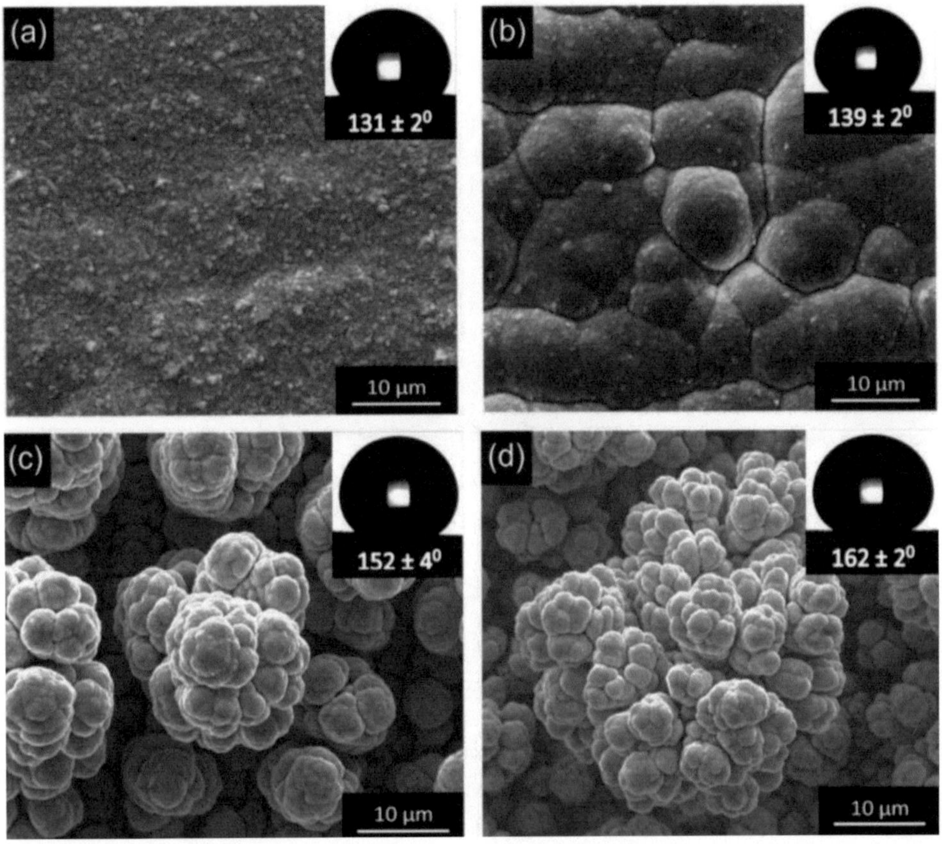

Figure 22. SEM images of surface morphology and WCA are shown at different values of voltages including 0.5 V, 0.7 V, 0.9 V, and 1.1 V [71].

0.6 A cm^{-2} for 20 min. The result was a stainless-steel mesh with hierarchical micro- and nanoscale roughness as shown in **Figure 24**. The WCA was 162° and OCA[3] was 0°.

In **Figure 25**, a schematic of the final substrate after electrodeposition is shown. The Cu and Ni molecules are co-deposited by the conjugated pDOp-NDM. As mentioned before to achieve superhydrophobicity, hierarchical surface structure and low surface energy are needed at the same time. In this case, the pDOp-NDM molecules drastically reduced surface energy. Besides, the pDOp acts as a bonding agent which increases the bonding of Cu and Ni to the surface and also attaches them to the NDM. At last, the superhydrophobic and superoleophilic mesh had a high separation efficiency, good recyclability, and strong durability [72].

Su et al. [73] fabricated a robust abrasion and corrosion-resistant superhydrophobic coating on copper substrate by electrodeposition method. The nickel electrodeposition in this study was obtained through Watts bath consisting of $NiSO_4$, $NiCl_2.6H_2O$, and H_3BO_3. The electrodeposition current density was 0.75 A cm^{-2} and the duration time was 1 h. After that, the sample was

[3]Oil contact angle.

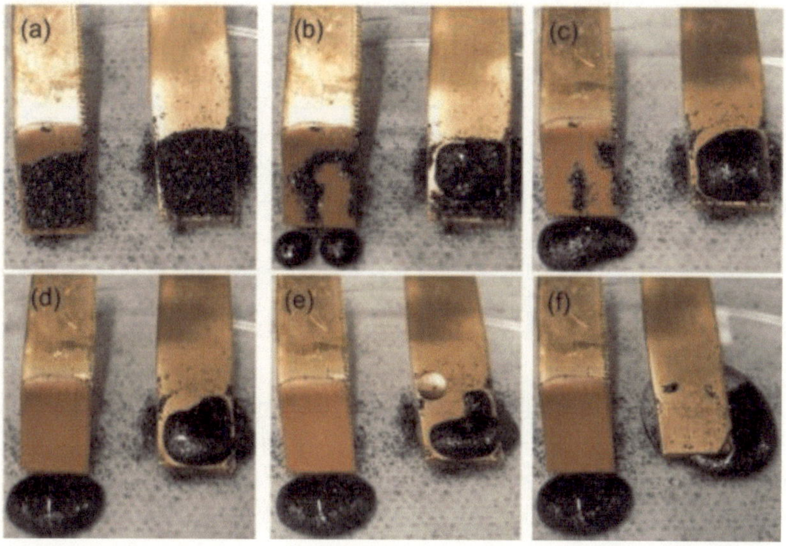

Figure 23. Self-cleaning properties of superhydrophobic copper substrate fabricated by electrodeposition method and compared to the ordinary substrate (a) dirty samples by SiC particles, (b) 2, (c) 5, (d) 10, (e) 30, and (f) 55 drops of water.

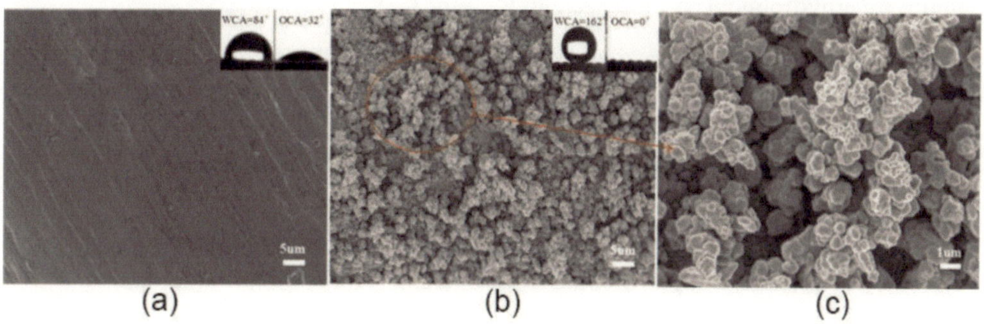

Figure 24. WCA and OCA on stainless steel substrate before (a) and after (b and c) electrodeposition process [72].

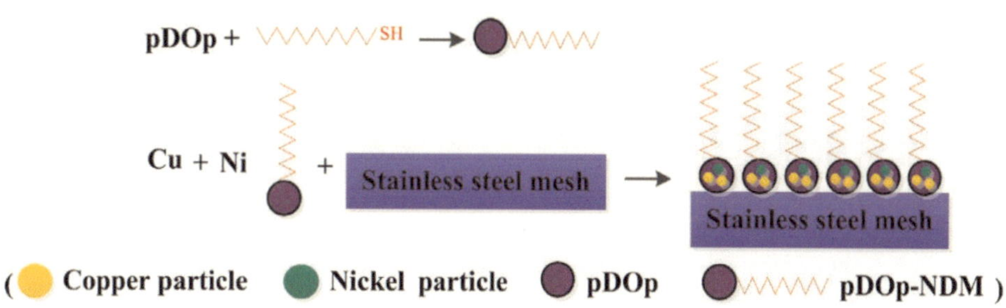

Figure 25. The conjugated pDOp-NDM anchored on Cu and nucleus [72].

put into a sealed reactor containing AC-FAS ethanol solution for 1 h at 110°C. In **Figure 26**, the schematic of the coating process is shown.

The nickel coating on top provided a hierarchical structure on surface, and AC-FAS lowered the surface energy resulted in superhydrophobic properties with good mechanical properties and high corrosion resistance. In **Figure 27**, Nyquist and bode plots of pure copper substrate, electrodeposited Ni, and superhydrophobic surface are shown in which the superhydrophobic surface has much higher corrosion resistance [73].

5.1.5. Chemical functionalization

Wu et al. [74] fabricated a ZnO-based surface on a glass slide in which a layer of ZnO was first deposited on the substrate, and then self-assembled monolayers[4] were used to lower surface energy and achieve superhydrophobic properties. In **Figure 28**, the microstructure of the ZnO coating on the substrate is shown. The hierarchical micro- and nanoroughness of the ZnO coating followed by low surface energy treatment will lead to superhydrophobic properties.

Xu et al. [74] fabricated a superhydrophobic nanocomposite TiO_2/polystyrene coating which can be deposited through simple spray coating. TiO_2 nanoparticles were first functionalized by PFOA and then added to the polystyrene matrix and deposited on the substrate by spraying. It was found that usage of equal amounts of functionalized TiO_2 and polystyrene led into optimum superhydrophobic properties with WCA of 166°.

Wang et al. [74] fabricated a superhydrophobic surface on PDMS using modified ZnO particles. The ZnO particles were fabricated through a CVD process, and then they were functionalized to reduce the chance of agglomeration. In **Figure 29**, as-prepared ZnO rods are shown which are greatly suitable to form hierarchical micro- and nanoscale roughness. Also, coating WCA and microstructure are shown.

Wu et al. [75] fabricated a superhydrophobic coating by simply adding functionalized silica nanoparticles. The silica nanoparticles are fist added to an ethanol solution containing PTES to

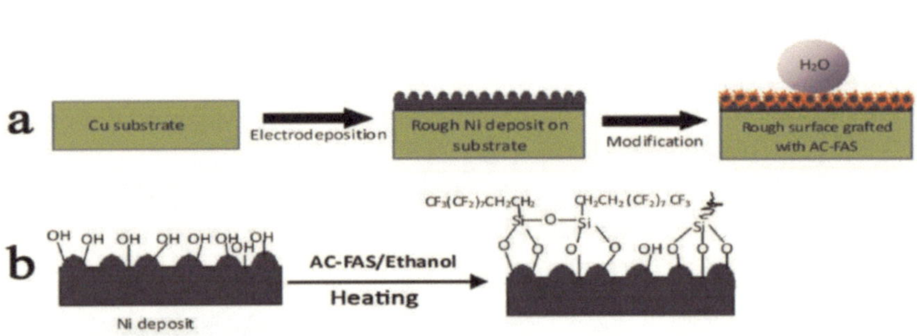

Figure 26. Schematics of Cu substrate electrodeposited by Ni followed by AC-FAS treatment [73].

[4]SAMs.

form hydrophobic silica particles, and then they were added to the epoxy to form the nanocomposite coating. As shown in **Figure 30**, the coating deposition of the substrate is not limited to only one method, and it can be brushed, dipped, and sprayed to the substrate.

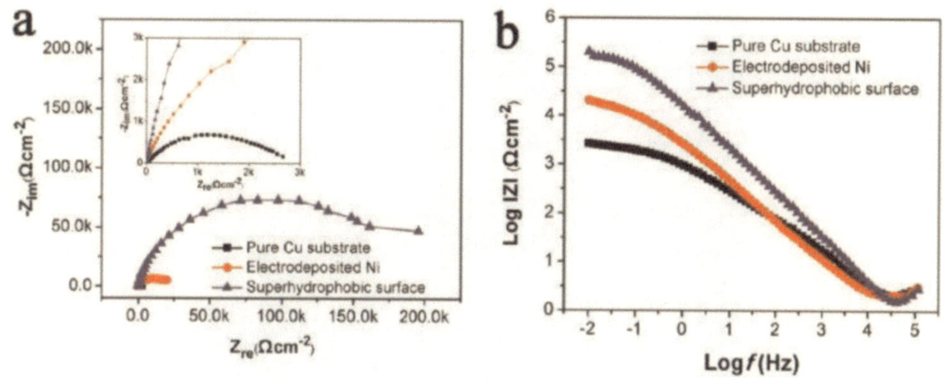

Figure 27. Nyquist and bode plots of pure Cu substrate, electrodeposited Ni, and superhydrophobic surface [73].

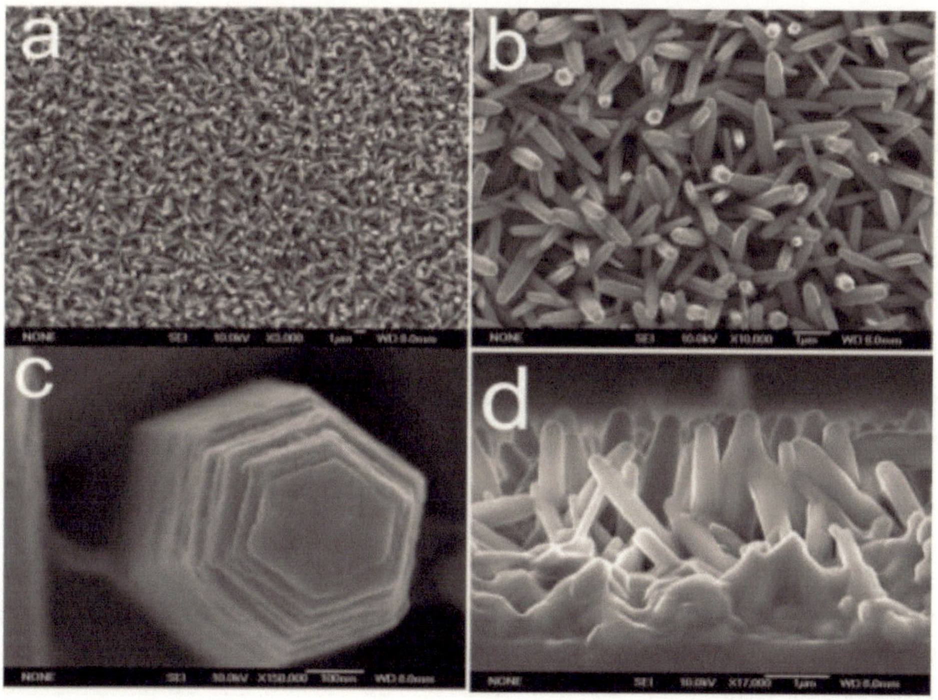

Figure 28. FESEM image of the surface at (a) low magnification, (b) high magnification, (c) hexagonal nanorod, and (d) cross-sectional view [74].

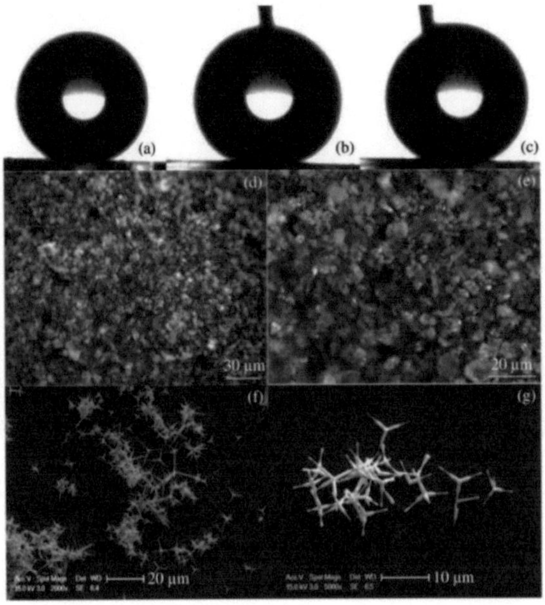

Figure 29. (a, b, c) WCA, advancing and receding angles, (d, e) coating microstructure, (f and g) as-prepared ZnO rods [74].

5.2. Lithography

The lithography method is one of the well-known processes used to fabricate superhydrophobic coatings. This method includes light-assisted, soft, nano-, electron beam-assisted, X-ray, and colloidal lithography. In this method, superhydrophobic surface is generally made using a soft or hard surface as a reference and makes a new surface by copying the reference surface [76]. Different methods of lithography are not independent of each other, and it is possible two use two methods of lithography during the fabrication process. For example, light-assisted lithography is used during nanolithography [77].

Before in traditional lithography, a smooth surface was used as a reference, but now in the photolithography process, all the details of surface roughness can be identified by the use of X-ray and a photosensitive thin film. In soft lithography also an elastomer material is used for molding the surface, and then that is used for templating from the surface [78].

5.3. Templating

Templating method is one of the methods to fabricate superhydrophobic surfaces in which usually lithography method is used to build a template from the reference surface, and then that template will be used for templating procedure. The template surface can be a paper filter, insect wing, some animal's skin, and plant leaf. From the chemical and morphological aspect, the template can even be a molecule or polymer [79, 80].

In general, this method includes making a template and molding from that and then building a surface by using that mold (see **Figure 31**). One of the ways to build a superhydrophobic

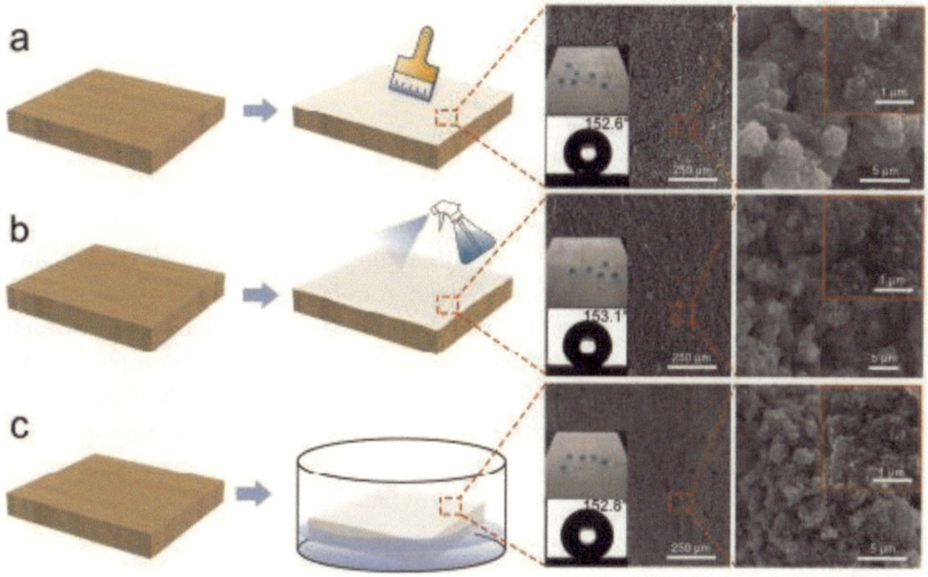

Figure 30. Superhydrophobic nanocomposite coating deposited on substrate through different methods and their microstructure and WCA measurements [75].

surface is to use gecko's feet as a template [81]. In **Figure 31**, the templating procedure from the lotus leaf is shown; in this procedure, first the lotus leaf is fixed on a prepared holder, and this leaf is coated with gold, and a nickel mold is used to build a mold from the lotus leaf.

5.4. Chemical vapor deposition

This method includes deposition of gas on substrate by chemical reactions. To evaporate the material in the CVD process, plasma, laser, catalysts, etc. can be used to ease the process. Some works that have used this method to achieve superhydrophobic properties will be mentioned here. Borras et al. [83] used plasma-assisted CVD to achieve superhydrophobicity by Ag-TiO_2 nanofibers on the surface of the substrate. The fibers included an Ag core wrapped with a TiO_2 shell. The water contact angle of the surface depended on the shape of the fibers and space between them. In the best situation, WCA was reached to near 180°. In superhydrophobic coatings with TiO_2 and ZnO, a change in wetting properties from superhydrophobic to superhydrophilic through UV irradiation can be observed. The reason for this unexpected change in wetting behavior is that UV irradiation will cause electron-hole pairs on the surface which will react with the lattice oxygen, and this will cause oxygen vacancy on the surface. These oxygen vacancies cause the water molecules to easily bond with the surface, and as a result, WCA will be drastically reduced to near zero.

In another study, Jung and Bhushan [84] fabricated multi-walled CNT by catalyst-assisted CVD and then combined that with resin and sprayed it on the Si substrate with microstructural roughness. The WCA of the surface was 170° and the CAH was about 2°. The superhydrophobic properties were stable in this sample and showed good results after long exposure to water.

Many other studies have used CVD to fabricate superhydrophobic coating, but the process is very complicated, and it is not possible to control the resulting surface morphology completely.

5.5. Layer-by-layer deposition

This method is an easier method than CVD and plasma which does not need very special equipment. In this method, several layers of thin coating will be applied on the substrate by changing the electrical charge, and there is less limitation in the size of the sample than CVD. Layer-by-layer deposition is a relatively easy method to fabricate a hierarchical structure on the substrate. Usually, some nanoparticle additives are used to improve the surface roughness. Cohen and Rubner's group [85, 86] used this method to fabricate superhydrophobic coatings. They used different kinds of solutions to fabricate three layers of the coating including adhesion, body, and top layer. In **Figure 32**, the schematic of the LBL process shows three solutions, consisting of polyallylamine hydrochloride[5], sodium 4-styrenesulfonate[6], and silica particles.

5.6. Colloidal aggregation and assembling

This method is usually used to ease sol-gel, chemical deposition, and lithography processes to fabricate hierarchical micro- and nanoscale roughness. In this method, chemical and van der Waals bond between the dispersed particles and deposition of these particles on the surface will make a multilayer rough surface [87]. To fabricate these kinds of multilayer rough surface, immersion or spin coating is used. For example, Min et al. [88] fabricated colloidal particles with a controlled size of 70 nm which were possible to deposit on large surfaces using a spin coating method. In **Figure 33a**, the TEM image of synthesized silica nanoparticles is shown. In **Figure 33b**, the SEM image of coating surface structure after deposition on the substrate by spin coating with 10,000 rpm and for 10 minutes is shown. The thickness of the crystalline colloidal layer can be controlled by the speed and time of the spin coating process.

5.7. Electrospinning and electrospraying

These two methods are similar to each other and are used to fabricate micro- or nanostructures. The electrospinning method is an easy way to fabricate continuous polymer fibers in micro- and nanoscale [89]. This method is suitable to make a roughened surface needed to achieve superhydrophobicity on the surface [90]. To produce uniform fibers, polymer molecular weight and solution concentration should be controlled [91].

On the other hand, the electrospraying method is not just limited to fibers, and the deposited polymer film can have different shapes from spheres to fibers [92]. Generally, polymeric fibers are produced through electrospinning, and films consist of spherical seeds, which are fabricated by electrospraying method [93].

---

[5]PAH.

[6]SPS.

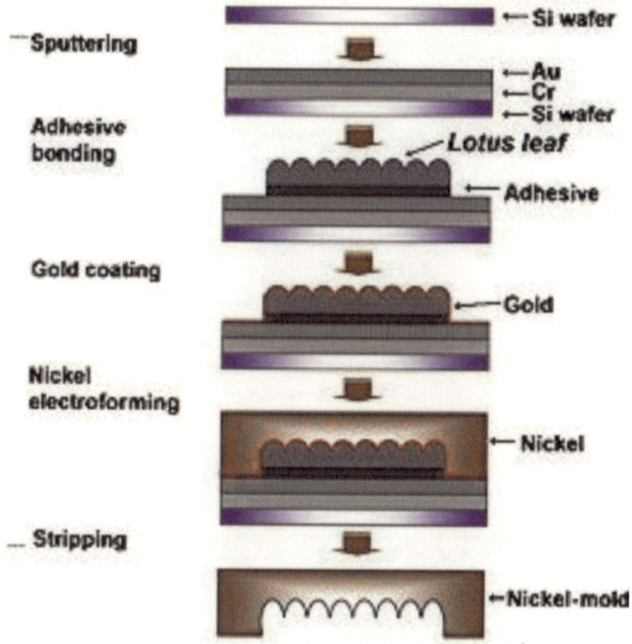

Figure 31. Schematics of the templating procedure from lotus leaf [82].

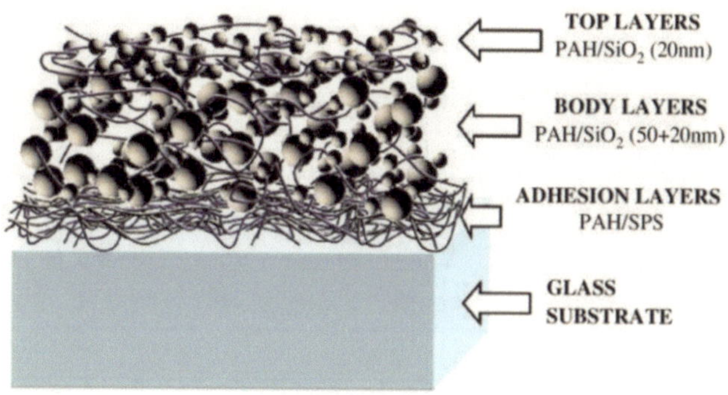

Figure 32. Schematic of the LBL method used to fabricate superhydrophobic coating [85].

Ding et al. [94] fabricated superhydrophobic nanostructured ZnO coating by the electrospinning method. The composite coating consisted of polyvinyl alcohol[7] and ZnO nanofibers. After that to reduce surface energy, fluoroalkyl silane was used to modify the surface and lower the energy on the surface, and superhydrophobicity was achieved. The FESEM image of inorganic ZnO fibers is shown in **Figure 34**. Before surface modification with

---

[7]PVA.

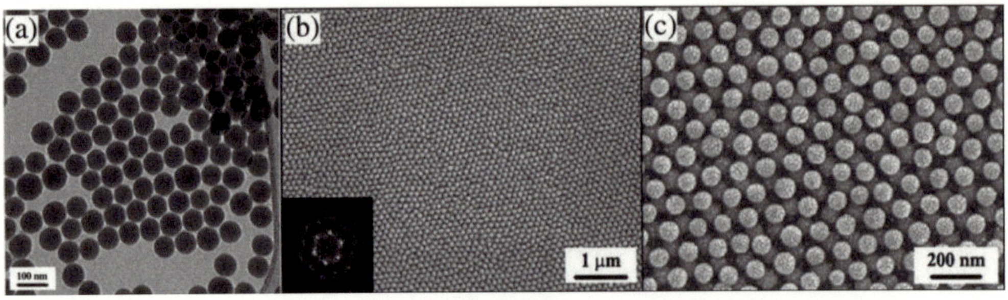

Figure 33. (a) TEM image of synthesized silica nanoparticles, (b, c) SEM image of the coating surface structure on substrate [88].

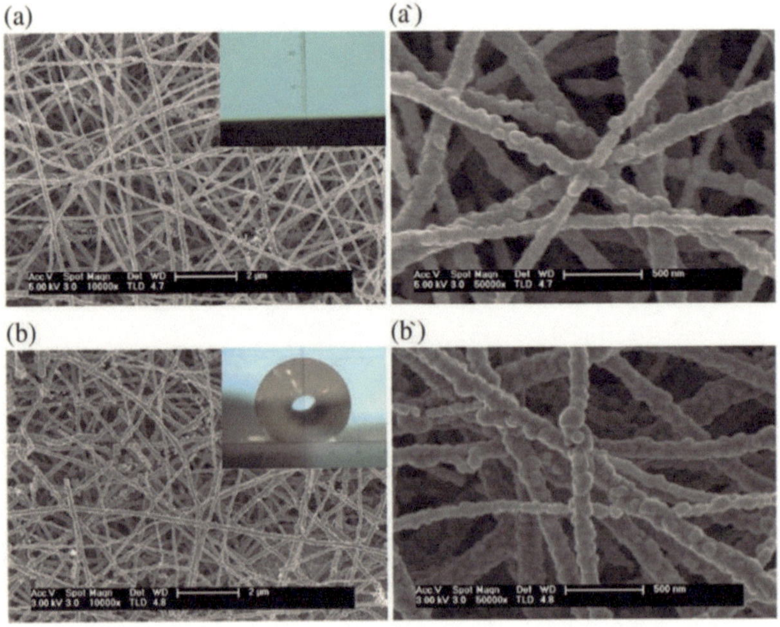

Figure 34. FESEM image of ZnO fibers film (a, a') before modification and (b, b') after modification [94].

fluoroalkyl silane, WCA was 0°, and the surface was superhydrophilic, but after surface modification, the WCA was 165°, and the sliding angle was 5°.

Burkarter et al. [95] fabricated a PTFE film by electrospray method on a glass substrate which had a fluorine-doped tin oxide coating. The result was a superhydrophobic coating with WCA equal to 160° and a sliding angle less than 2°. Actually, the electrospray method used in this study was very similar to the electrospinning, but because there was no need for PTFE fiber, then the electrospray method was used. The SEM image of hierarchical micro- and nanoscale roughness of the coating is shown in **Figure 35**.

Figure 35. SEM image of hierarchical micro- and nanoscale roughness of PTFE film deposited by electrospray method on a glass substrate which had a fluorine-doped tin oxide coating [95].

5.8. Phase separation

In this method, a surface pattern is made by separation solid phase from a semi-stable mixture by changing the temperature, pressure, or other environmental conditions. Phase separation method can also be involved in colloidal assembling. The surface structure in this method can be in macro-, micro-, or nanoscale [96]. This method is usually followed by sol-gel method to control resulted surface pattern [97]. Phase separation method can also be followed by other methods like plasma treatment, electrospinning, and self-aggregation to achieve superhydrophobicity. The phase separation method is partially related to colloidal polymerization [98].

6. Drawbacks of superhydrophobic coatings

There have been many improvements in the case of superhydrophobic coating fabrication and their stability, but there is still room to grow. Many studies on the superhydrophobic coatings do not consider mechanical properties as their focus or at least part of the study. Many of superhydrophobic coatings have poor bonding either to the substrate or to itself. Also, many other superhydrophobic coatings will lose their special wetting behavior during long-term use or in harsh environments. Also, almost in most superhydrophobic coatings, a low surface energy treatment is done, which does not have suitable bonding and stability on coating substrate and, in the case of fluorine-based materials, is toxic and harmful for environments. These weak points have hindered industrial application of superhydrophobic coatings.

Author details

Sepehr Shadmani[1], Mehdi Khodaei[1]*, Xiuyong Chen[2,3] and Hua Li[2,3]

*Address all correspondence to: khodaei@kntu.ac.ir

1 Faculty of Materials Science and Engineering, K.N. Toosi University of Technology, Tehran, Iran

2 Key Laboratory of Marine Materials and Related Technologies, Zhejiang Key Laboratory of Marine Materials and Protective Technologies, Ningbo Institute of Materials Technology and Engineering, Chinese Academy of Sciences, Ningbo, China

3 Cixi Institute of Biomedical Engineering, Ningbo Institute of Materials Technology and Engineering, Chinese Academy of Sciences, Ningbo, China

References

[1] Goodwyn PP, Maezono Y, Hosoda N, Fujisaki K. Waterproof and translucent wings at the same time: Problems and solutions in butterflies. Naturwissenschaften. 2009;**96**:781-787

[2] Gao X, Jiang L. Water-repellent legs of water striders. Nature. 2004;**432**:36-36

[3] Das S, Kumar S, Samal SK, Mohanty S, Nayak SK. A review on superhydrophobic polymer nanocoatings: Recent development and applications. Industrial & Engineering Chemistry Research. 2018;**57**:2727-2745

[4] Gao L, McCarthy TJ. Contact angle hysteresis explained. Langmuir. 2006;**22**:6234-6237

[5] Hao P, Lv C, Yao Z, He F. Sliding behavior of water droplet on superhydrophobic surface. Europhysics Letters. 2010;**90**:66003

[6] Wang B, Liang W, Guo Z, Liu W. Biomimetic super-lyophobic and super-lyophilic materials applied for oil/water separation: A new strategy beyond nature. Chemical Society Reviews. 2015;**44**:336-361

[7] Gao C, Sun Z, Li K, Chen Y, Cao Y, Zhang S, et al. Integrated oil separation and water purification by a double-layer TiO_2-based mesh. Energy & Environmental Science. 2013;**6**: 1147-1151

[8] Crick CR, Gibbins JA, Parkin IP. Superhydrophobic polymer-coated copper-mesh; membranes for highly efficient oil–water separation. Journal of Materials Chemistry A. 2013;**1**: 5943-5948

[9] Nguyen DD, Tai N-H, Lee S-B, Kuo W-S. Superhydrophobic and superoleophilic properties of graphene-based sponges fabricated using a facile dip coating method. Energy & Environmental Science. 2012;**5**:7908-7912

[10] Zhu Q, Chu Y, Wang Z, Chen N, Lin L, Liu F, et al. Robust superhydrophobic polyurethane sponge as a highly reusable oil-absorption material. Journal of Materials Chemistry A. 2013;**1**:5386-5393

[11] Zhang Y-L, Xia H, Kim E, Sun H-B. Recent developments in superhydrophobic surfaces with unique structural and functional properties. Soft Matter. 2012;**8**:11217-11231

[12] Fan Y, Li C, Chen Z, Chen H. Study on fabrication of the superhydrophobic sol-gel films based on copper wafer and its anti-corrosive properties. Applied Surface Science. 2012; **258**:6531-6536

[13] Ishizaki T, Sakamoto M. Facile formation of biomimetic color-tuned superhydrophobic magnesium alloy with corrosion resistance. Langmuir. 2011;**27**:2375-2381

[14] Xu W, Song J, Sun J, Lu Y, Yu Z. Rapid fabrication of large-area, corrosion-resistant superhydrophobic Mg alloy surfaces. ACS Applied Materials & Interfaces. 2011;**3**:4404-4414

[15] Tian Y, Su B, Jiang L. Interfacial material system exhibiting superwettability. Advanced Materials. 2014;**26**:6872-6897

[16] de Leon ACC, Pernites RB, Advincula RC. Superhydrophobic colloidally textured polythiophene film as superior anticorrosion coating. ACS Applied Materials & Interfaces. 2012;**4**:3169-3176

[17] Wang N, Xiong D. Superhydrophobic membranes on metal substrate and their corrosion protection in different corrosive media. Applied Surface Science. 2014;**305**:603-608

[18] Gnedenkov S, Egorkin V, Sinebryukhov S, Vyaliy I, Pashinin A, Emelyanenko A, et al. Formation and electrochemical properties of the superhydrophobic nanocomposite coating on PEO pretreated Mg-Mn-Ce magnesium alloy. Surface and Coatings Technology. 2013;**232**:240-246

[19] Cheng J, Zhang J, Liu F. Recent development of metal hydroxides as electrode material of electrochemical capacitors. RSC Advances. 2014;**4**:38893-38917

[20] Chen X, Yuan J, Huang J, Ren K, Liu Y, Lu S, et al. Large-scale fabrication of superhydrophobic polyurethane/nano-Al_2O_3 coatings by suspension flame spraying for anti-corrosion applications. Applied Surface Science. 2014;**311**:864-869

[21] Wang G, Guo Z, Liu W. Interfacial effects of superhydrophobic plant surfaces: A review. Journal of Bionic Engineering. 2014;**11**:325-345

[22] Tung WS, Daoud WA. Self-cleaning fibers via nanotechnology: A virtual reality. Journal of Materials Chemistry. 2011;**21**:7858-7869

[23] Gao L, McCarthy TJ. Wetting 101°. Langmuir. 2009;**25**:14105-14115

[24] Ho AYY, Luong Van E, Lim CT, Natarajan S, Elmouelhi N, Low HY, et al. Lotus bioinspired superhydrophobic, self-cleaning surfaces from hierarchically assembled templates. Journal of Polymer Science Part B: Polymer Physics. 2014;**52**:603-609

[25] Daoud WA, editor. Self-Cleaning Materials and Surfaces: A Nanotechnology Approach. NYC, USA: John Wiley & Sons; Jul 12 2013

[26] Wisdom KM, Watson JA, Qu X, Liu F, Watson GS, Chen C-H. Self-cleaning of superhydrophobic surfaces by self-propelled jumping condensate. Proceedings of the National Academy of Sciences. 2013;**110**:7992-7997

[27] Meuler AJ, McKinley GH, Cohen RE. Exploiting topographical texture to impart icephobicity. ACS Nano. 2010;**4**:7048-7052

[28] Guo P, Zheng Y, Wen M, Song C, Lin Y, Jiang L. Icephobic/anti-icing properties of micro/nanostructured surfaces. Advanced Materials. 2012;**24**:2642-2648

[29] Peng C, Xing S, Yuan Z, Xiao J, Wang C, Zeng J. Preparation and anti-icing of superhydrophobic PVDF coating on a wind turbine blade. Applied Surface Science. 2012; **259**:764-768

[30] Boinovich L, Emelyanenko AM. Role of water vapor desublimation in the adhesion of an iced droplet to a superhydrophobic surface. Langmuir. 2014;**30**:12596-12601

[31] Li X-M, Reinhoudt D, Crego-Calama M. What do we need for a superhydrophobic surface? A review on the recent progress in the preparation of superhydrophobic surfaces. Chemical Society Reviews. 2007;**36**:1350-1368

[32] Richard D, Quéré D. Bouncing water drops. Europhysics Letters. 2000;**50**:769

[33] Wang Y, Xue J, Wang Q, Chen Q, Ding J. Verification of icephobic/anti-icing properties of a superhydrophobic surface. ACS Applied Materials & Interfaces. 2013;**5**:3370-3381

[34] Sohn Y, Kim D, Lee S, Yin M, Song JY, Hwang W, et al. Anti-frost coatings containing carbon nanotube composite with reliable thermal cyclic property. Journal of Materials Chemistry A. 2014;**2**:11465-11471

[35] Wang Y, Li M, Lv T, Wang Q, Chen Q, Ding J. Influence of different chemical modifications on the icephobic properties of superhydrophobic surfaces in a condensate environment. Journal of Materials Chemistry A. 2015;**3**:4967-4975

[36] Hejazi V, Sobolev K, Nosonovsky M. From superhydrophobicity to icephobicity: Forces and interaction analysis. Scientific Reports. 2013;**3**:2194

[37] McHale G, Newton MI, Shirtcliffe NJ. Immersed superhydrophobic surfaces: Gas exchange, slip and drag reduction properties. Soft Matter. 2010;**6**:714-719

[38] Bhushan B. Bioinspired structured surfaces. Langmuir. 2012;**28**:1698-1714

[39] Su B, Li M, Lu Q. Toward understanding whether superhydrophobic surfaces can really decrease fluidic friction drag. Langmuir. 2009;**26**:6048-6052

[40] Watanabe K, Udagawa Y, Udagawa H. Drag reduction of Newtonian fluid in a circular pipe with a highly water-repellent wall. Journal of Fluid Mechanics. 1999;**381**:225-238

[41] Moaven K, Rad M, Taeibi-Rahni M. Experimental investigation of viscous drag reduction of superhydrophobic nano-coating in laminar and turbulent flows. Experimental Thermal and Fluid Science. 2013;**51**:239-243

[42] Deng Z-Y, Wang W, Mao L-H, Wang C-F, Chen S. Versatile superhydrophobic and photocatalytic films generated from TiO_2–SiO_2@ PDMS and their applications on fabrics. Journal of Materials Chemistry A. 2014;**2**:4178-4184

[43] Dong H, Cheng M, Zhang Y, Wei H, Shi F. Extraordinary drag-reducing effect of a superhydrophobic coating on a macroscopic model ship at high speed. Journal of Materials Chemistry A. 2013;**1**:5886-5891

[44] Rothstein JP. Slip on superhydrophobic surfaces. Annual Review of Fluid Mechanics. 2010;**42**:89-109

[45] Banerjee I, Pangule RC, Kane RS. Antifouling coatings: Recent developments in the design of surfaces that prevent fouling by proteins, bacteria, and marine organisms. Advanced Materials. 2011;**23**:690-718

[46] Schierholz J, Beuth J. Implant infections: A haven for opportunistic bacteria. Journal of Hospital Infection. 2001;**49**:87-93

[47] Hetrick EM, Schoenfisch MH. Reducing implant-related infections: Active release strategies. Chemical Society Reviews. 2006;**35**:780-789

[48] Guo Z, Zhou F, Hao J, Liu W. Stable biomimetic super-hydrophobic engineering materials. Journal of the American Chemical Society. 2005;**127**:15670-15671

[49] Heinonen S, Huttunen-Saarivirta E, Nikkanen J-P, Raulio M, Priha O, Laakso J, et al. Antibacterial properties and chemical stability of superhydrophobic silver-containing surface produced by sol–gel route. Colloids and Surfaces A: Physicochemical and Engineering Aspects. 2014;**453**:149-161

[50] Xue C-H, Chen J, Yin W, Jia S-T, Ma J-Z. Superhydrophobic conductive textiles with antibacterial property by coating fibers with silver nanoparticles. Applied Surface Science. 2012;**258**:2468-2472

[51] Song J, Xu W, Liu X, Lu Y, Wei Z, Wu L. Ultrafast fabrication of rough structures required by superhydrophobic surfaces on Al substrates using an immersion method. Chemical Engineering Journal. 2012;**211**:143-152

[52] Ma W, Wu H, Higaki Y, Otsuka H, Takahara A. A "non-sticky" superhydrophobic surface prepared by self-assembly of fluoroalkyl phosphonic acid on a hierarchically micro/nanostructured alumina gel film. Chemical Communications. 2012;**48**:6824-6826

[53] Mahadik SA, Kavale MS, Mukherjee S, Rao AV. Transparent superhydrophobic silica coatings on glass by sol–gel method. Applied Surface Science. 2010;**257**:333-339

[54] Kim E-K, Lee C-S, Kim SS. Superhydrophobicity of electrospray-synthesized fluorinated silica layers. Journal of Colloid and Interface Science. 2012;**368**:599-602

[55] Brassard J-D, Sarkar DK, Perron J. Fluorine based superhydrophobic coatings. Applied Sciences. 2012;**2**:453-464

[56] Brassard J-D, Sarkar DK, Perron J. Synthesis of monodisperse fluorinated silica nanoparticles and their superhydrophobic thin films. ACS Applied Materials & Interfaces. 2011;**3**:3583-3588

[57] Huang Y, Sarkar DK, Chen X-G. A one-step process to engineer superhydrophobic copper surfaces. Materials Letters. 2010;**64**:2722-2724

[58] Safaee A, Sarkar DK, Farzaneh M. Superhydrophobic properties of silver-coated films on copper surface by galvanic exchange reaction. Applied Surface Science. 2008;**254**:2493-2498

[59] Saleema N, Sarkar DK, Gallant D, Paynter RW, Chen X-G. Chemical nature of superhydrophobic aluminum alloy surfaces produced via a one-step process using fluoroalkyl-silane in a base medium. ACS Applied Materials & Interfaces. 2011;**3**:4775-4781

[60] Sarkar DK, Farzaneh M, Paynter RW. Superhydrophobic properties of ultrathin rf-sputtered Teflon films coated etched aluminum surfaces. Materials Letters. 2008;**62**:1226-1229

[61] Sarkar DK, Saleema N. One-step fabrication process of superhydrophobic green coatings. Surface and Coatings Technology. 2010;**204**:2483-2486

[62] Teshima K, Sugimura H, Inoue Y, Takai O. Gas barrier performance of surface-modified silica films with grafted organosilane molecules. Langmuir. 2003;**19**:8331-8334

[63] Zhao Y, Li M, Lu Q, Shi Z. Superhydrophobic polyimide films with a hierarchical topography: Combined replica molding and layer-by-layer assembly. Langmuir. 2008;**24**: 12651-12657

[64] Hozumi A, Takai O. Effect of hydrolysis groups in fluoro-alkyl silanes on water repellency of transparent two-layer hard-coatings. Applied Surface Science. 1996;**103**:431-441

[65] Latthe SS, Imai H, Ganesan V, Rao AV. Superhydrophobic silica films by sol–gel co-precursor method. Applied Surface Science. 2009;**256**:217-222

[66] Khodaei M, Shadmani S. Superhydrophobicity on aluminum through reactive-etching and TEOS/GPTMS/nano-Al_2O_3 silane-based nanocomposite coating. Surface and Coatings Technology. 2019;**374**:1078-1090

[67] Gu C, Zhang T-Y. Electrochemical synthesis of silver polyhedrons and dendritic films with superhydrophobic surfaces. Langmuir. 2008;**24**:12010-12016

[68] Xu X, Zhang Z, Yang J. Fabrication of biomimetic superhydrophobic surface on engineering materials by a simple electroless galvanic deposition method. Langmuir. 2009;**26**:3654-3658

[69] Haghdoost A, Pitchumani R. Fabricating superhydrophobic surfaces via a two-step electrodeposition technique. Langmuir. 2014;**30**:4183-4191

[70] Zhang W, Yu Z, Chen Z, Li M. Preparation of super-hydrophobic Cu/Ni coating with micro-nano hierarchical structure. Materials Letters. 2012;**67**:327-330

[71] Jain R, Pitchumani R. Facile fabrication of durable copper-based superhydrophobic surfaces via electrodeposition. Langmuir. 2017;**34**:3159-3169

[72] Xiang M, Jiang M, Zhang Y, Liu Y, Shen F, Yang G, et al. Fabrication of a novel superhydrophobic and superoleophilic surface by one-step electrodeposition method for continuous oil/water separation. Applied Surface Science. 2018;**434**:1015-1020

[73] Su F, Yao K. Facile fabrication of superhydrophobic surface with excellent mechanical abrasion and corrosion resistance on copper substrate by a novel method. ACS Applied Materials & Interfaces. 2014;**6**:8762-8770

[74] Wu X, Zheng L, Wu D. Fabrication of superhydrophobic surfaces from microstructured ZnO-based surfaces via a wet-chemical route. Langmuir. 2005;**21**:2665-2667

[75] Wu Y, Jia S, Wang S, Qing Y, Yan N, Wang Q, et al. A facile and novel emulsion for efficient and convenient fabrication of durable superhydrophobic materials. Chemical Engineering Journal. 2017;**328**:186-196

[76] Kwon Y, Patankar N, Choi J, Lee J. Design of surface hierarchy for extreme hydrophobicity. Langmuir. 2009;**25**:6129-6136

[77] Choi Y-W, Han J-E, Lee S, Sohn D. Preparation of a superhydrophobic film with UV imprinting technology. Macromolecular Research. 2009;**17**:821-824

[78] Zheng B, Tice JD, Ismagilov RF. Formation of arrayed droplets by soft lithography and two-phase fluid flow, and application in protein crystallization. Advanced Materials. 2004;**16**:1365-1368

[79] Lai Y, Huang Y, Wang H, Huang J, Chen Z, Lin C. Selective formation of ordered arrays of octacalcium phosphate ribbons on TiO_2 nanotube surface by template-assisted electrodeposition. Colloids and Surfaces B: Biointerfaces. 2010;**76**:117-122

[80] Bormashenko E, Stein T, Whyman G, Bormashenko Y, Pogreb R. Wetting properties of the multiscaled nanostructured polymer and metallic superhydrophobic surfaces. Langmuir. 2006;**22**:9982-9985

[81] Cho WK, Choi IS. Fabrication of hairy polymeric films inspired by geckos: Wetting and high adhesion properties. Advanced Functional Materials. 2008;**18**:1089-1096

[82] Yan Y, Gao N, Barthlott W. Mimicking natural superhydrophobic surfaces and grasping the wetting process: A review on recent progress in preparing superhydrophobic surfaces. Advances in Colloid and Interface Science. 2011;**169**:80-105

[83] Borras A, Barranco A, González-Elipe AR. Reversible superhydrophobic to superhydrophilic conversion of Ag@TiO_2 composite nanofiber surfaces. Langmuir. 2008;**24**:8021-8026

[84] Jung YC, Bhushan B. Mechanically durable carbon nanotube-composite hierarchical structures with superhydrophobicity, self-cleaning, and low-drag. ACS Nano. 2009;**3**:4155-4163

[85] Bravo J, Zhai L, Wu Z, Cohen RE, Rubner MF. Transparent superhydrophobic films based on silica nanoparticles. Langmuir. 2007;**23**:7293-7298

[86] Lee D, Rubner MF, Cohen RE. All-nanoparticle thin-film coatings. Nano Letters. 2006;**6**: 2305-2312

[87] Sun C, Ge L-Q, Gu Z-Z. Fabrication of super-hydrophobic film with dual-size roughness by silica sphere assembly. Thin Solid Films. 2007;**515**:4686-4690

[88] Min W-L, Jiang P, Jiang B. Large-scale assembly of colloidal nanoparticles and fabrication of periodic subwavelength structures. Nanotechnology. 2008;**19**:475604

[89] Ma M, Hill RM, Lowery JL, Fridrikh SV, Rutledge GC. Electrospun poly (styrene-block-dimethylsiloxane) block copolymer fibers exhibiting superhydrophobicity. Langmuir. 2005;**21**:5549-5554

[90] Tuteja A, Choi W, Ma M, Mabry JM, Mazzella SA, Rutledge GC, et al. Designing superoleophobic surfaces. Science. 2007;**318**:1618-1622

[91] Ma M, Mao Y, Gupta M, Gleason KK, Rutledge GC. Superhydrophobic fabrics produced by electrospinning and chemical vapor deposition. Macromolecules. 2005;**38**:9742-9748

[92] Mizukoshi T, Matsumoto H, Minagawa M, Tanioka A. Control over wettability of textured surfaces by electrospray deposition. Journal of Applied Polymer Science. 2007;**103**:3811-3817

[93] Zheng J, He A, Li J, Xu J, Han CC. Studies on the controlled morphology and wettability of polystyrene surfaces by electrospinning or electrospraying. Polymer. 2006;**47**:7095-7102

[94] Ding B, Ogawa T, Kim J, Fujimoto K, Shiratori S. Fabrication of a super-hydrophobic nanofibrous zinc oxide film surface by electrospinning. Thin Solid Films. 2008;**516**:2495-2501

[95] Burkarter E, Saul CK, Thomazi F, Cruz NC, Zanata SM, Roman LS, et al. Electrosprayed superhydrophobic PTFE: A non-contaminating surface. Journal of Physics D: Applied Physics. 2007;**40**:7778

[96] Li X, Chen G, Ma Y, Feng L, Zhao H, Jiang L, et al. Preparation of a super-hydrophobic poly (vinyl chloride) surface via solvent–nonsolvent coating. Polymer. 2006;**47**:506-509

[97] Nakajima A, Abe K, Hashimoto K, Watanabe T. Preparation of hard super-hydrophobic films with visible light transmission. Thin Solid Films. 2000;**376**:140-143

[98] Shirtcliffe N, McHale G, Newton M, Perry C. Intrinsically superhydrophobic organosilica sol-gel foams. Langmuir. 2003;**19**:5626-5631